SpringerBriefs in Applied Sciences and Technology

Nanoscience and Nanotechnology

Series editor

Hilmi Volkan Demir, Nanyang Technological University, Singapore, Singapore

Nanoscience and nanotechnology offer means to assemble and study superstructures, composed of nanocomponents such as nanocrystals and biomolecules, exhibiting interesting unique properties. Also, nanoscience and nanotechnology enable ways to make and explore design-based artificial structures that do not exist in nature such as metamaterials and metasurfaces. Furthermore, nanoscience and nanotechnology allow us to make and understand tightly confined quasi-zero-dimensional to two-dimensional quantum structures such as nanopalettes and graphene with unique electronic structures. For example, today by using a biomolecular linker, one can assemble crystalline nanoparticles and nanowires into complex surfaces or composite structures with new electronic and optical properties. The unique properties of these superstructures result from the chemical composition and physical arrangement of such nanocomponents (e.g., semiconductor nanocrystals, metal nanoparticles, and biomolecules). Interactions between these elements (donor and acceptor) may further enhance such properties of the resulting hybrid superstructures. One of the important mechanisms is excitonics (enabled through energy transfer of exciton-exciton coupling) and another one is plasmonics (enabled by plasmon-exciton coupling). Also, in such nanoengineered structures, the light-material interactions at the nanoscale can be modified and enhanced, giving rise to nanophotonic effects.

These emerging topics of energy transfer, plasmonics, metastructuring and the like have now reached a level of wide-scale use and popularity that they are no longer the topics of a specialist, but now span the interests of all "end-users" of the new findings in these topics including those parties in biology, medicine, materials science and engineerings. Many technical books and reports have been published on individual topics in the specialized fields, and the existing literature have been typically written in a specialized manner for those in the field of interest (e.g., for only the physicists, only the chemists, etc.). However, currently there is no brief series available, which covers these topics in a way uniting all fields of interest including physics, chemistry, material science, biology, medicine, engineering, and the others.

The proposed new series in "Nanoscience and Nanotechnology" uniquely supports this cross-sectional platform spanning all of these fields. The proposed briefs series is intended to target a diverse readership and to serve as an important reference for both the specialized and general audience. This is not possible to achieve under the series of an engineering field (for example, electrical engineering) or under the series of a technical field (for example, physics and applied physics), which would have been very intimidating for biologists, medical doctors, materials scientists, etc.

The Briefs in NANOSCIENCE AND NANOTECHNOLOGY thus offers a great potential by itself, which will be interesting both for the specialists and the non-specialists.

More information about this series at http://www.springer.com/series/11713

Won Kook Choi

ZnO–Nanocarbon Core–Shell Type Hybrid Quantum Dots

 Springer

Won Kook Choi
Materials and Life Science Research
  Division
Korea Institute of Science and Technology
Seoul
Korea, Republic of (South Korea)

ISSN 2191-530X   ISSN 2191-5318   (electronic)
SpringerBriefs in Applied Sciences and Technology
ISSN 2196-1670   ISSN 2196-1689   (electronic)
Nanoscience and Nanotechnology
ISBN 978-981-10-0979-2   ISBN 978-981-10-0980-8   (eBook)
DOI 10.1007/978-981-10-0980-8

Library of Congress Control Number: 2016947379

Printed on acid-free paper

This Springer imprint is published by Springer Nature
The registered company is Springer Science+Business Media Singapore Pte Ltd.

Preface

Lower-cost and higher-performing metal oxides have emerged as alternatives to precious noble and rare materials due to abundance, stability, and electrical and chemical capacities. In particular, enormous interests have been paid on metal oxide/nanocarbons (CNT, C_{60}, graphene, carbon fiber, amorphous carbon, etc.) hybrid materials in the electrochemical reactions on energy conversion and storage and their electrocatalyst systems. It has been demonstrated that assembling metal or metal oxide nanomaterials (Au, Pt, TiO_2, ZnO, SnO_2, etc.) on graphene sheets can exhibit enhanced efficiencies in excitonic solar cells and photocatalytic reactions, due to graphene's excellent electron-conducting property. Also, various kinds of metal oxide with CNT have been used as electrocatalysts for water splitting, as supercapacitors, as anodes for Li ion secondary battery anode, as sensors detecting reducing gases, and as charge transfer layers for solar photovoltaic cells and light-emitting diodes. These hybrid materials showed prominently enhanced performance because nanocarbons have flexibility, easy functionalization, and high electrical and thermal conductivity.

Hybridization of metal oxide with nanocarbons was known to easily occur from chemical reaction between metal oxide and induced chemical groups on nanocarbon surfaces by facile acid treatment. And now the strongly coupled metal oxide/nanocarbon hybrid can be synthesized by controlling the optimum degree of oxidation, and thereby balancing inorganic carbon coupling showing high electrical performance. Despite many efforts to manipulate more uniform metal oxide–nanocarbon nanocomposite structures, the metal oxide nanoparticles were still randomly scattered and nonuniformly attached to the nanocarbon surfaces. Also, in most cases the hybrid materials still remain 2D planar structures. For higher and more effective performance of the hybrid structure, 3D conformal coating on metal oxides is highly demanded. To our knowledge, consolidated core–shell structure metal oxide NPs encircled by nanocarbons with high conformality have been rarely reported. Since consolidated ZnO–graphene core–shell type hybrid quantum dots were for the first time synthesized using chemical reaction between ZnO embryo nanoparticle and acid-treated graphene in 2012, this emissive hybrid quantum dots were used for the realization of white light-emitting ZnO–graphene quantum dots (ZGQDs) LED. In this book, ZnO–templated synthetic method to form conformal 3D nanocarbon hybrid materials is introduced.

In Chap. 1, as an introduction, the general physical and chemical properties of ZnO and nanocarbons (CNT, graphene, C_{60}) are briefly summarized. As one of the important applications of metal oxide nanoparticles, recent research on charge transport layer or charge injection layer of ZnO or TiO_2 or their composites with polymer is overall reviewed from the former studies in electronic devices of solar photovoltaic cells, electrochemical electrodes, and light-emitting diodes. In addition, previous researches on metal oxide–nanocarbon hybrid structures are also introduced in the fields of supercapacitors, Li ion secondary battery, electrocatalysts, photovoltaic cells, and light-emitting diodes.

In Chap. 2, synthetic processes of ZnO–graphene and ZnO–C_{60} hybrid quantum dots, and the formation of nanoring single-walled CNT using ZnO–SW (single-walled) CNT are described in detail. Nanostructure of these hybrid quantum dots are precisely analyzed by high-resolution TEM, HR-HAADF (high-resolution high-angle annular dark field) STEM (scanning transmission electron spectroscopy), and X-ray diffraction. Chemical functional group induced at the interface between ZnO nanoparticles and nanocarbon surface is confirmed by X-ray photoelectron spectroscopy. Optical properties of these hybrid quantum dots are investigated by Raman spectroscopy, which is specifically well known for analyzing nanocarbon materials. The charge transfer phenomenon from the conduction band of ZnO to graphene quantum dots is carefully examined by the fitting curves of the time-resolved photoluminescence (TRPL) decay (the biexponential function calculates the lifetime in the UV range). Based on density functional theory (DFT), two blue emissions in PL newly arising in ZnO–graphene QDs are well explained as electron transitions from the conduction band (CB) of ZnO, lowest unoccupied molecular orbital (LUMO), LUMO+2 levels induced by epoxy oxygen on graphene to the valence band (VB) of ZnO. The formation of nanoring SWCNT, with the diameter of 20–30 nm, is well described by the agglomeration of ZnO nanoparticles.

In Chap. 3, applications of these ZnO–graphene, ZnO–C_{60} hybrid quantum dots, and NR-SWCNT are introduced. In the cases of ZnO–graphene and ZnO–C_{60} hybrid quantum dots, four things are presented: UV photovoltaic solar cells, high-efficiency inverted ZnO–graphene QD-based white LED, flexible QD LED, and in a photoelectrochemical water splitting device, the high-performing photoanode of ZnO–C_{60}. Also, as an example of NR-SWCNT application, when P(VDF-TrFE) piezoelectric polymer is mixed with NR-SWCNT, an enormous increase of permittivity from $\varepsilon = 10$–12 to ca. 63 is observed during a small dielectric tangent loss (tan δ)—as much as merely 0.06 at 1 kHz.

I would like to appreciate all members of the Soft Nano Electronic Laboratory (SNEL)—Mr. Dong-Hee Park, Mr. Se-Hee Cho, Mr. Tae-Hee Yoo, Mr. Choong Hye Kim, Mr. Jung-Hyuk Kim, Mr. Ju-Won Lim, Mr. Chang-Gui Jin, Mr. Bum Hee Lee, Mr. Chang-Hwan Wie, Dr. Rina Pandey, and Dr. Young-Taek Lee—for their continuous and persistent collaboration; in particular, my colleague Dr. D.I. Son, for his vigorous research for a number of applications of ZnO–nanocarbon hybrid materials; Mr. Byong-Wook Kwon, for the first realization of synthesis of the ZnO–graphene hybrid quantum dot structure; Dr. Do-Kyung Hwang, Dr. Young-Soo No, Dr. Jeong-Do Yang, and Mr. Hong-Hee Kim, for fabricating the flexible ZnO–graphene QD LED; Dr. S. Bae, for his contribution to fabricate and analyze the solar PVs using modified ZnO–graphene QDs; Prof. J.H. Park of Yonsei University, for his contribution to developing photoelectrochemical cells using ZnO–graphene as well as ZnO–C_{60}; Mr. Yun-Jae Lee, Ms. So-Ra Ham, and Prof. S.R. Kim, for developing and synthesizing the nanoring single-walled CNT; both Dr. Won-Seon Seo of KICET—for the transmission electron microscope analysis and valuable comments—and Dr. Chang-Lyoul Lee of GIST—for TRPL measurement and interpretation of the spectra, respectively; Prof. Yeonjin Yi and his graduate students, for the DFT simulation and calculation of the ZnO–graphene hybrid material; Dr. B. Angadi, Bangalore University, for his valuable comments and advice in composition; and Mrs. Cindy Zitter of Springer SBM NL and Mr. Smith Chae of Springer Korea, for their advice and guidance in the publication of this book.

Special thanks to my family: Eunice, Daniel, and Sue.

Seoul, Korea, Republic of (South Korea)   Won Kook Choi

Contents

Chapter 1
Introduction

1.1 Zinc Oxide (ZnO)

ZnO is one of typical nonstoichometric oxides and also of the most abundant materials in the earth and no hazardous to human, and eco-friendly material. Because of deficiency of oxygen (oxygen vacancy), it has been used prototype gas sensor material as like SnO_2 by using the change of surface conductivity as reducing gases of CH_4, C_3H_8 etc. were adsorbed and desorbed on the surface (Kolmakov et al. 2003; Chou et al. 2006; Wang et al. 2010).

Since the late 1990s, ZnO has been explosively emerged as new candidate for highly efficient and brilliant near ultraviolet (NUV) light emitting diode (LED) and exponentially investigated until mid 2000s. Epitaxial growth, p-type doping, multi quantum well (MQW) structure, dilute magnetic semiconductor (DMS) at room temperature were main research topics in ZnO material. Electronically it is a semiconducting oxide material with large bandgap 3.37 eV at room temperature. In particular, since it has large exciton binding energy as much as 60 meV (Look 2001) at room temperature quite larger than 28 meV of GaN and 20 meV of ZnS, it was highly expected for near UV exciton-emitting emitter at room temperature and photodetector. Moreover, high oscillation strength defined as $f = 2P^2/m(h/2\pi)\omega$, $P = \langle c/H_{eR}/v \rangle$, $f_{ex}/f = (E_b/E)^{3/2}$, it is expected to be promising candidate for high excitonic optical device. Moreover, it shows higher optical gain of 300 cm^{-1} than 100 cm^{-1} (Hvam 1978) of GaN and also high saturation velocity of 3.2×10^7 m/s (Solis-Pomar et al. 2011) nearly equivalent to or higher than GaN, InGaN, and AlGaN (Morkoc et al. 1994; Anwar et al. 2001). ZnO has a high melting point ($T_m = 2248$ K), and decomposes rapidly at temperatures above 1650 K, which is

© The Author(s) 2017
W.K. Choi, *ZnO–Nanocarbon Core–Shell Type Hybrid Quantum Dots*,
Nanoscience and Nanotechnology, DOI 10.1007/978-981-10-0980-8_1

difficult to achieve bulk crystal growth (Klimm et al. 2011). ZnO single crystal has been grown by (a) hydrothermal (Liu and Zeng 2003; Guo et al. 2005; Baruah and Dutta 2009) and (b) seeded chemical vapor transport (SCVT) or sublimation method (Ntep et al. 1999; Yao and Hong 2009; Kohl et al. 1974) (c) the Bridgman method (Schulz et al. 2006) and the other (Berry and Deutschman Jr. 1971) up to 6 inch in diameter. In Table 1.1, general physical properties of ZnO are summarized.

Existence of ZnO homo crystal means the possibility and get over the most difficult obstacle in the growth of sound GaN epitaxial film on hetero substrate of c-plane sapphire single crystal substrate with large misfit of ca. 16 % without buffer layer like AlN. Highly crystalline ZnO crystal could be grown at lower temperature around 550 °C by rf magnetron sputtering or 600 °C by reactive plasma assisted molecular bam epitaxy (MBE), or 700 °C by reactive pulsed laser deposition which were all quite lower growth temperature than 900 °C for GaN growth. Therefore ZnO has been extensively investigated as the most promising candidate of active material emitting blue or near UV light replacing III-V's GaN based blue LED. In its application to optoelectronic devices, the main bottleneck lies in the lack of reproducible epitaxial p-type ZnO growth with high conductivity. Besides various physical properties of ZnO were well described for the material for blue/UV optoelectronics, including light-emitting or even laser diodes in addition to (or instead of) the GaN-based structures, a radiation hard material for electronic devices in a corresponding environment, a material for electronic circuits, which is transparent in the visible, a diluted or ferromagnetic material, when doped with Co, Mn, Fe, V or similar elements, for semiconductor spintronics, and a transparent, highly conducting oxide (TCO), when doped with Al, Ga, In or similar elements, as a cheaper alternative to indium tin oxide (ITO) (Özgür et al. 2005; Klingshirn 2007).

After 2006, considerable current interest for ZnO research seemed to be converted from epitaxial layers and quantum wells to nanostructures such as nanorods, nanobelt, nanowire-related objects or quantum dots. Since ZnO could be easily formulized into nanostructure, these semiconducting oxide material were fascinating for exploiting very efficient energy harvesting, thin film transistor (TFT), nanorod type LED, oxide based logic devices, and gas sensor etc. including analyzing synthetic mechanism. But how well understanding and control of intrinsic defects of ZnO still remains as a crucial point for achieving prominent ZnO-based devices.

Table 1.1 Physical constants of ZnO

Crystal structure	Lattice constants		Band-gap E_g@RT	Melting point, T_m	Exciton binding energy	Di-electric constant $\varepsilon(0)$	Thermal conductivity, W
	a(Å)	c(Å)					
Wurtzite	3.248	5.209	3.37 (eV)	2250 (K)	60 (meV)	8.75	0.6 (cm K)

1.2 ZnO Nanoparticles in Electronic Devices

Oxide semiconducting nanoparticles have been widely adopted as carrier transport materials in energy and environmental electronic, electrochemical, and photoelectrochemical devices such as electro photovoltaic cells, light emitting diodes, secondary batteries, photoelectrochemical electrodes etc.

Among them, titanium oxide (TiO_2) and zinc oxide (ZnO) have most extensively used. TiO_2 has usually two crystalline phase of anatase and rutile. The former structure has been known to be crystallized at lower temperature below 450 °C than that of latter case. On the other hand, ZnO nanoparticle could be synthesized relatively lower temperature than that for TiO_2 even below 150 °C. In future electronics, wearable devices which should be endurable and strong lifetime at the severe conditions like bending, folding, and stretching will be highly expected and overwhelmed. Accordingly, low temperature fabrication process will be highly demanded so that devices should be constructed on the soft and flexible substrates such as polymer materials. In relation with flexible electronics, ZnO will be most promising materials to meet such requirement due to low crystallization temperature, no hazardous, and abundance.

In organic material based devices like OLED (organic light emitting diode) and OPVs (organic photovoltaic cells) device, degradation resulting from exposure to oxygen/moisture and thermal instability are main drawbacks. In order to solve these problems, inorganic layers instead of organic materials were suggested to be replaced for improving the device performance. For instance, p-type NiO and n-type ZnO as hole transport layer (HTL) and electron transport layer (ETL) respectively were adopted in QD LED. As one example, a TiO_2 sol-gel precursor (DuPont tyzol BTP) diluted to 5 wt% butanol was spin-coated on CdSe/CdS/ZnS quantum dot emitting layer and annealed at low temperature (<150 °C). Low temperature annealing could not fully crystallize the transport TiO_2 layer. From the ultraviolet photoemission spectroscopy (UPS) and UV-Vis absorption spectrum, the valence band maximum (VBM), the conduction band maximum (CBM), and the fermi level were measured as 7.8, 3.9, and 4.56 eV respectively. In TiO_2 ETL-based QD LED, it was Al/TiO_2 led to increase electron-injection current and decrease the turn-on voltage (1.9 V) much lower than that (4.0 eV) for Alq_3-based device, which is attributed to both higher electron mobility (1.7×10^{-4} cm^2/V s) (Kim et al. 2006) than that ($\sim 1 \times 10^{-5}$ cm^2/V s) of Alq_3 (Kepler et al. 1995) and reduced band offset of Al/TiO_2 (0.4 eV) compared to Al/Alq_3 (1.2 eV). It was revealed that insertion of low temperature annealed and not fully crystalline TiO_2 ETL was very effective in electron injection and finally superior to that for Alq_3 ETL device. This device showed a high red luminance (12,380 cd/m^2) and high power efficiency (2.41 lm/W) (Cho et al. 2009).

Polycrystalline ZnO nanoparticles with the size of ~ 3 nm were synthesized by solution precipitation method mixture with both Zn acetate in dimethyl sulphoxide (DMSO) and tetramethylammonium hydroxide (TMAH) in ethanol (Tan et al. 2007). Solution-processed ZnO nanoparticles (25–75 nm) were adopted as electron

transport layer (ETL) in all solution processed CdSe–ZnS quantum dot LED
sandwiched by poly-TPD organic hole transport layer (HTL) on ITO/PEDOT:PSS
(AI4803) anode (Qian et al. 2011). Compared to the electroluminescent QD LED
devices without ETL (Al-only) or with a conventional Alq$_3$/Al, lowering of operational voltage lighting brightness level was mainly explained by the increment of
electron current density. These were ascribed to both higher electron mobility with
an order of 10^{-2}–10^{-3} cm^2/V s of ZnO nanoparticle than that ca. 1×10^{-4} cm^2/V s
of organic HTL and efficient electron injection into QD by Auger process. Electron
injection into the QDs from ZnO NPs was enhanced by well adjustment of energy
level of conduction band maximum (4.3 eV) with the Fermi level of Al and large
accumulation of electrons at the interface of poly-TPD/QD at which the Auger
injections effectively occurred. In consequence, if TiO$_2$ ETL was replaced by more
high mobility ZnO (2×10^{-3} to 0.07 cm^2/V s) ETL, the devices had maximum
luminance and power efficiency values of 4200 cd/m^2 and 0.17 lm/W for blue
emission, 68,000 cd/m^2 and 8.2 lm/W for green, and 31,000 cd/m^2 and 3.8 lm/W
for orange-red emission.

In previous (Qian et al. 2010), carrier recombination at polymer/ZnO interface
were systematically studied. In case of hole-dominant polymer (MEH-PPV,
MDMO-PPV)/ZnO, turn-on voltage (V$_{th}$) was quite lower than and electroluminescence (EL) started to be observed at drive voltages below the corresponding
bandgap voltage (V$_{ph}$ = hv/e) of emitting polymer. This kind of sub-band EL could
be attributed to an Auger-assisted energy up-conversion at the MEH-PPV/ZnO
nanoparticles (NPs) interface as schematically represented in Fig. 1.1.

At forward bias, both hole injected from PEDOT:PSS/MEH-PPP and electrons
from ZnO/Al were accumulated at the MEH-PPV/ZnO interface. These electrons and
holes form interfacial charge transfer (CT) excitons, or exciplex states. Photons
released from the recombination of these CT excitons are nonradiatively absorbed by
electrons on the ZnO NPs which is called Auger process. These high energy Auger

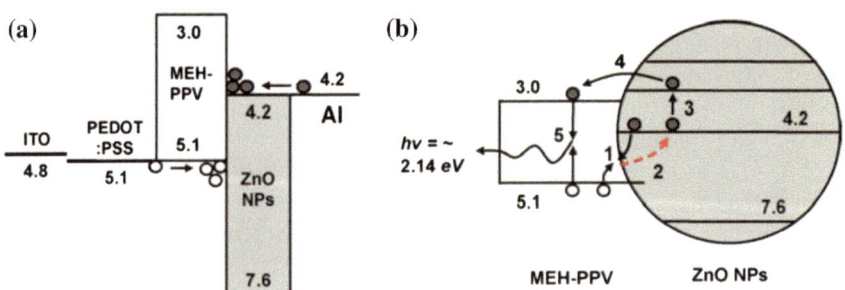

Fig. 1.1 **a** Energy band diagram of MEH-PPV and ZnO NPs interface. **b** Schematic diagram of
the Auger-like energy up-conversion process at the interface of MEH-PPV and ZnO NPs. (*1*)
recombination of interfacial charge transfer (*CT*) exciton, (*2*) resonant energy transfer between
exciton and electron, (*3*) Auger electron generation by energy transfer, (*4*) charge injection from
CB of ZnO NPs to MEH-PPV LUMO, and (*5*) radiation from electron-hole recombination in
MEH-PPV (Reproduced from Qian et al. 2010)

electrons are injected into the energy level of lowest unoccupied molecular orbital (LUMO) of the MEH-PPV polymer by electric fields at voltages well below the bandgap and then radiatively recombine with holes in the highest occupied molecular orbital (HOMO). It is noteworthy that the V_{th} becomes small as the size of ZnO NPs decreases. Such a dependence of V_{th} on the dimension of ZnO NPs indicates that Auger up-conversion process is related to the quantum confinement of the ZnO NPs and correlates to how much the CT excitons overlap with electron wave functions. On the other hand, accumulation of holes at the electron-dominant (PFO, F8BT) or ambipolar (THB) polymer/ZnO NPs interface is not effectively occurred and thus measured V_{th} are always larger than those of V_{ph} due to insufficient Auger process. In a similar ways, ZnO NPs have been used as ETL in conventional QD LED with Al (Ho et al. 2013; Kim et al. 2015a, b). As above mentioned, difference of electrical mobility of hole in transport material usually invokes the imbalance of the number of between hole and electron carriers at active layer. To optimize charge balance in the normal structure, a very thin insulating poly(methylmethacrylate) (PMMA, 6 nm) was inserted between thin-shell quantum dots emitter and ZnO NPs/Al cathode. Because the electron mobility of ZnO (1.8×10^{-3} cm^2/V s) was measured one or three orders of magnitude higher than the hole mobility of poly-TPD (1×10^{-4} cm^2/V s) and PVK (2.5×10^{-6} cm^2/V s) (Thesen et al. 2010). A device consisting of ITO/PEDOT:PSS/poly-TPD/PVK/QDs/ZnO/Ag yields a peak external quantum efficiency (EQE) of 2.5 % and shows increment of EQE up to 4.7 % through adding 6 nm-thick PMMA layer (Dai et al. 2014).

In terms of formation of organic hole transport layer in conventional OLED, minimizing the physical damage of HTL during solution process is another critical issue to be overcome to achieve high performance and endurable devices. As a solution, an inverted structure LED was suggested in which inorganic semiconducting oxide NPs have been adequately used for electron injection (transport) layer (EIL, ETL) instead of organic layer (Kwak et al. 2012) (Fig. 1.2a) reported that ZnO NPs spin-coated on ITO as a cathode showed high transmittance in the visible range and high electron mobility as much as $\mu_e \sim 1.3 \times 10^{-3}$ cm^2/V s and didn't allow the quenching of exciton formed at QDs interface rather than the other metal oxides prepared by sputtering or thermal annealing. This inverted QD LED showed high blue, green, and red brightness of maximum luminance s up to 2250, 218,800, and 23,040 cd/m^2, and external quantum efficiency 1.7, 5.8, and 7.3 % respectively In addition to brightness, low turn-on voltage, high efficiency, and operational stability, inverted structure LQD has advantages of utilizing low-price n-channel thin-film transistor (TFT) backplane which allows direct programming of the TFT gate-source voltage. Recently it was also revealed that the adjacent ZnO NPs ETL layer was electronically coupled with adjacent quantum dots emitting layer and took charge of both electron transfer and maintaining charge balance which was largely dependent on the thickness of QDs layer (Fig. 1.2b) (Mashford et al. 2013).

In inverted polymer light-emitting diode (PLED), the LUMOs of most polymer emitting layers (PEL) are (2.8–3.2 eV) (Kabra et al. 2010) shallower than that of ZnO and thus the high energy injection barrier into PLE should be lowered to achieve highly efficient PLED. For this, polymeric high dipole layers such as

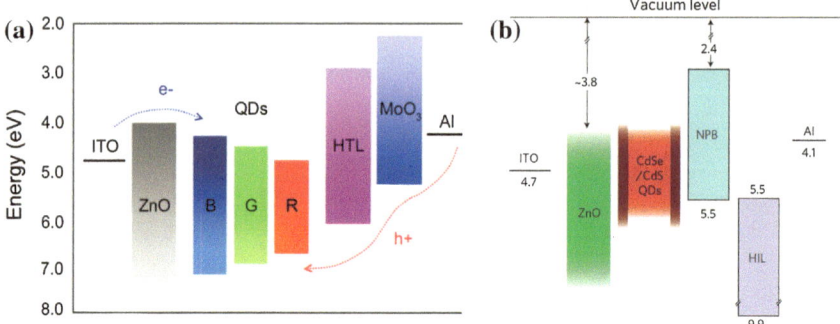

Fig. 1.2 Flat band diagram of inverted QLED structure with efficient full-color QDs. **a** *red* QDs, **b** at no bias (Reproduced from Kwak et al. 2012; Mashford et al. 2013)

polyethylenimine (PEI) (Chen et al. 2010), polyethylenimine ethoxylated (PEIE) (Zhou et al. 2012), and poly [(9.9-bis)3-(N,N-dimethylamino)prophyl]-2,7-fluorene)-alt-2,7-(9,9-dioctylfluorene)] were widely adopted. According to Kim et al. (Fig. 1.3a) the changes of work function (4.4 eV) of ZnO with no interlayer, 3.56 eV for ZnO/Cs$_2$CO$_3$, 3.29–3.6 eV (ZnO/PEIE(4, 16 nm)), and 2.47–3.39 eV (ZnO/PEI(4, 16 nm)) with the variations of thickness from 4 to 16 nm were accurately measured by ultraviolet spectroscopy (UPS) (Kim et al. 2014). This interlayer (or surface modifier) of PEI and PEIE were easily protonated and induced electrostatic dipole on ZnO surface. These polymer dipole played electron donor and therefore reduce the work function of ZnO which enhanced electron injection into LUMO of PEL. Moreover, these polymer dipole contribute also to improve efficiently the blocking of holes and of exciton quenching at ZnO/PEL interface. Similar approach based on ZnO/PEI was also reported (Hofle et al. 2014). Very recently, it was reported that the work function (3.58 eV) of ZnO NPs was largely reduced to 2.87 eV of ZnO/PEIE(7 nm) which was smaller than formerly reported 3.29–3.6 eV (Kim et al. 2014). This ZnO/PEIE layer was adopted in inverted CdSe/ZnS QLED and produced maximum luminance and current efficiency of 8600 cd/m^2 and 1.53 cd/A respectively (Fig. 1.3b) (Kim et al. 2015a, b). In p-DTS (FBTTH$_2$)$_2$, PC70BM bulk heterojunction solar cell, ZnO/PEIE was also used as interfacial layer. This efficiently reduced both work function and the trap assisted recombination and led to get the photoconversion efficiency of 7.88 % and internal quantum efficiency (IQE) almost equivalent to 100 % around 480 and 600 nm (Kyaw et al. 2013).

Xu et al. studied an efficient CsSe/CdS/ZnS QLED with three different TiO$_2$ (45 nm) with relatively lower electron mobility ($\sim 1.7 \times 10^{-4}$ cm^2/V s), TiO$_2$ (20 nm)/ZnO (25 nm) bilayer, and ZnO (45 nm) with relatively higher electron mobility ETLs. Through the device with TiO$_2$/ZnO bilayer ETL, they demonstrated there was a trade-off between luminance and efficiency. The QLED with ZnO showed high luminance, but low efficiency whereas the device with TiO$_2$ was vice versa. In the ZnO-based device, high mobility leads to QD charging and quenching

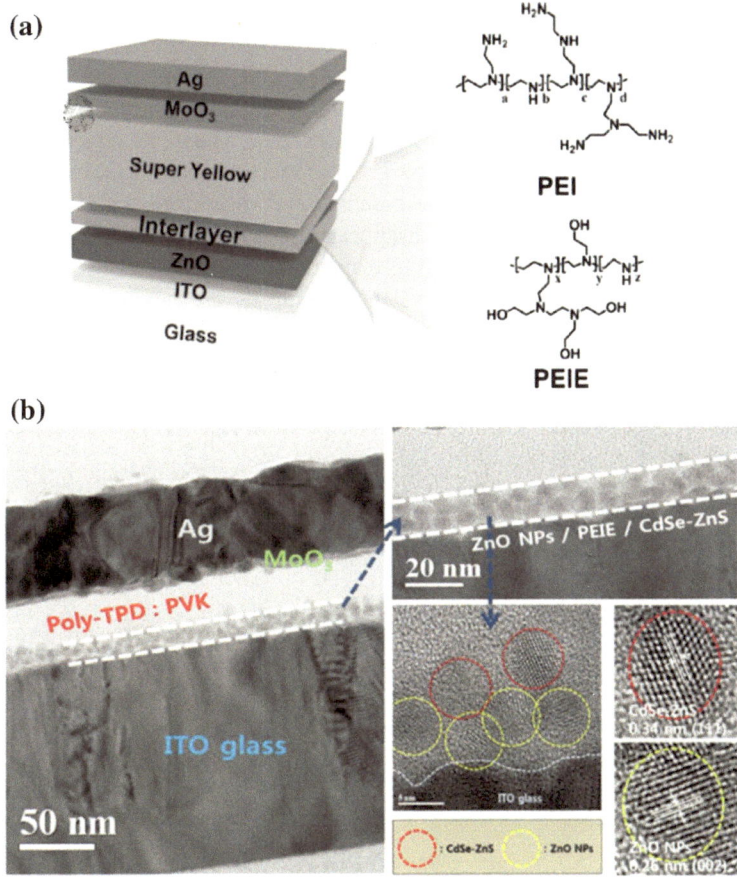

Fig. 1.3 **a** Inverted polymer LED structure using PEI and PEIE surface modifier on ZnO layer (Reproduced from Kim et al. 2014). **b** Cross sectional TEM image of ZnO/PEIE/red CdSe-ZnS LED (Reproduced from Kim et al. 2015a, b)

of exciton at QD. But in case of TiO_2-basd device, slow injection of electron can balance the charge injection into QD and leads to high current efficiency (cd/A) at high voltage and current density (mA/cm^2). In the ZnO-containing devices, electron injection occurs at low voltage and in parallel hole injection will be enhanced as the current density increases and thus decreases the of QD charging. Moreover, Auger assisted energy upconversion hole injection at CBP/QD interface at low current density would be enhanced by nonradiative photon emitted from recombination between injected electron from ZnO and hole accumulated at the CBP/QDs (Xu et al. 2014). Single hybrid $ZnO@TiO_2$ NPs electron transport layer was also tested in normal structured QLED of ITO/PEDOT:PSS/TFB/ZnCdSeS (590 nm)/ $ZnO@TiO_2$/Al where ZnO NPs played as both ETL/EIL and blue emitter. Considering the results revealed by Xu et al. (2014), exciton quenching at QD caused by high mobility ZnO ETL will be reduced by the adding rather lower mobility TiO$_2$ layer. Chen et al. (2012) obtained photoluminescence spectra (PL) as shown in Fig. 1.4 for ZnO NPs by scanning the excitation photon wavelength from 370 nm to 400 nm. They found that the green peak centered around 544 nm with blue peaks near at 434 and 463 nm was only observed when ZnO NPs were excited by the above bandgap energy and assigned as the transition of excited electrons from the conduction band into the deep defect level (Mashford et al. 2013). At the low bandgap excitation, green peak disappears and only blue peaks remains which are believed as the transition of electrons excited into Zn$_i$ ($\sim$2.9 eV) subbands (Zeng et al. 2010) and then returned to valence band. At the optimum thickness of PEDOT/TFS (20 nm), the peak intensity blue color electroluminescence was overwhelmed that of yellow peak from the QDs. At last white-light emitting QD LED with its initial luminance of 730 cd/m^2 was realized by the mixture of the emission form QDs and defect-related emission from ZnO NPs.

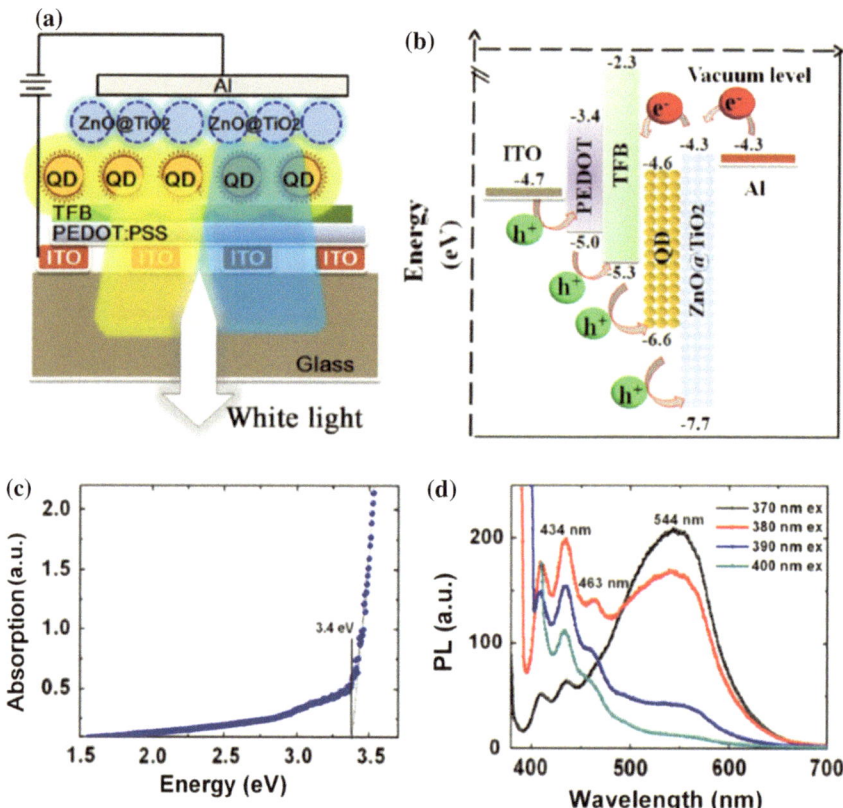

Fig. 1.4 **a** A schematic diagram of device structure and **b** the corresponding energy band diagram of QD LED with hybrid ZnO@TiO$_2$ film as electron injection layer. **c** UV-Vis absorption spectra and **d** PL spectra of hybrid ZnO@TiO$_2$ layer (Reproduced from Chen et al. 2014)

1.3 Nanocarbons

Carbon-carbon bonded materials may be existed in the several forms. Figure 1.5 illustrates eight different molecular configuration that carbon can take: graphite, diamond, lonsdaleite, C_{60} Buckminster fullerene (buckyball), C_{540} Fullerite, C_{70}, amorphous carbon (soot and charcoals), and single walled carbon nanotube (SW-CNT). Among them, C_{60}, CNT, and graphene are known as the allotropes with spherical, cylindrical, two-dimensional nanostrucutres respectively and typical nanocarbons.

Since Osawa predicted the existence of C_{60} in 1970 (Osawa 1970), it was firstly discovered by Harold Kroto in 1985 (Kroto et al. 1985) with the support of mass spectrometry. Spherical and cylindrical shaped C_{60}s are separately called as Buckminsterfullerene (buckyballs) resembling the soccer balls and carbon

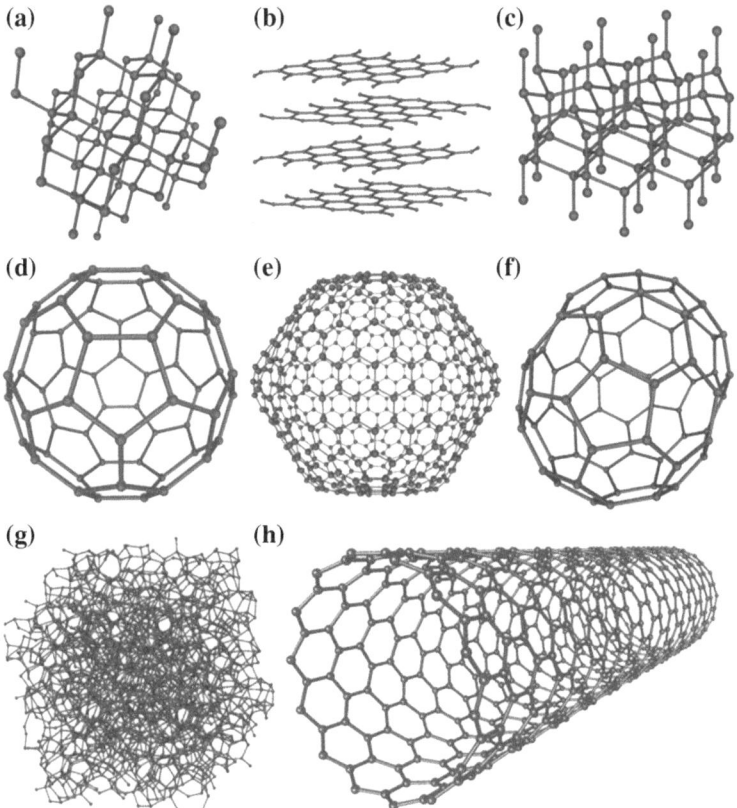

Fig. 1.5 Eight allotropes of carbon: **a** diamond, **b** graphite, **c** lonsdaleite, **d** C_{60} buckminster-fullerene, **e** C_{540}, Fullerite **f** C_{70}, **g** amorphous carbon, and **h** single-walled carbon nanotube (SW-CNT) (Reproduced from Wikipedia; https://commons.wikimedia.org/wiki/File%3AEight_Allotropes_of_Carbon.png)

nanotubes (buckytubes). In the amorphous carbon (soot), different form of C_{60}, C_{70}, C_{76}, C_{82} and C_{84} even up to C_{100} molecules were found formed by lightning discharges in the atmosphere (Dresselhaus 1995; Baena et al. 2002). The structure of a fullerene is described mathematically as a trivalent convex polyhedron with pentagonal and hexagonal faces, and C_{60} is made of 20 hexagons and 12 pentagons. The van der Waals diameter of a C_{60} molecule is about 1.1 nm (Qiao et al. 2007) and the nucleus to nucleus diameter of a C_{60} molecule is about 0.71 nm. The C_{60} molecule has two bond lengths: the 6:6 ring double bonds (between two hexagons) are shorter than the 6:5 bonds (between a hexagon and a pentagon). Its average bond length is 0.14 nm.

In nanotechnology, fullerene was studied as the materials of heat resistance and superconductivity. Fullerenes and their derivatives can play as a photosensitizer and thus have been extensively used for potential biomedical uses including the design of high-performance MRI contrast agents, X-Ray imaging contrast agents, photodynamic tumor therapy and drug and gene delivery, summarized in previous reviews (Tegos 2005; Lalwani and Sitharaman 2013). Recently PC$_{60}$ BM ([6,6]-phenyl-C(61)-butyric acid methyl ester) and its C_{70} derivatives (PC$_{70}$BM) as acceptor with low bandgap polymers as donors have been widely adopted in bulk heterojunction organic photovoltaic cells (He and Li 2010; Zhang et al. 2012).

Carbon nanotubes (CNTs) are members of the fullerene structural family as aforehand mentioned and allotropes of carbon with a cylindrical nanostructure and derived from the hollow structure wrapped with one-layer-carbon atom sheet, graphene. Nanotubes have been constructed with very high aspect ratio of length-to-diameter ratio of up to 132,000,000:1(Wang et al. 2009) significantly larger than for any other material. CNTs are wrapped with particular ("chiral") angles. Their electrically properties of metal or semiconductor are determined by the mixture of rolling angel and the radius. Nanotubes are divided as single-walled nanotubes (SWNTs) and multi-walled nanotubes (MWNTs). Each nanotubes naturally align themselves into "ropes" held together by van der Waals forces, more specifically, π-stacking. The chemical bonding of nanotubes is composed entirely of sp^2 bonds like graphite. Because of their extraordinary thermal conductivity and mechanical and electrical properties as enlisted in Table 1.2, these cylindrical carbon molecules have been extensively investigated for nanotechnology, electronics, optics and other fields of materials and biological science and technology such as file emission display (FED) CNT-based high speed transistor, CNT incorporated transparent electrode, and highly sensitive biological of DNA sensor (Li et al. 2003). In particular, carbon nanotubes have been used as additives or fillers to various structural hybrid functional materials. For instance, addition of a tiny portion of CNTs in some (primarily carbon fiber) baseball bats, golf clubs, car parts, thermal dissipation interface in high bright LED (Zhang et al. 2008) or damascus steel (Gullapalli and Wong 2011).

Graphene is an allotrope of carbon in the form of a 2-dim atomic-scale, honey-comb lattice in which one atom forms each vertex. It can also be considered as an indefinitely large aromatic molecule, the ultimate case of the family of flat polycyclic aromatic hydrocarbons (PAHs) in which interior carbons are surrounded

Table 1.2 Physical constants of crystalline Si, conventional representative materials, and nanocarbons (CNT and graphene)

	Crystalline Si	Related conventional materials	CNT	Graphene[a]
Mobility (@RT) (cm²/V s)	1500	0.1–1	1×10^5	2×10^5
Mean free path (@RT)			0.3–0.5	~ 1.0 μm
Resistivity (ρ) (Ωcm)	6.4×10^2		1.6×10^{-6}	1.0×10^{-6}
Bandgap (eV)	1.1–1.4		0.5–1.0	0–0.3
Thermal conductivity (W/mK)		380 (Cu) 900 (diamond)	3000–3500	5300[a]
Young's modulus (GPa)		220 (steel) 1200 (diamond)	1000–2000	1000
J_{max} (A/m²)		10^6 (Cu)	10^6	10^8
Surface area (m²/g)			~ 1500	~ 2630
Tensile strength (GPa)		<2 (steel)	30–180	10–20
Optical transmittance		88 % (ITO)		97.7 % $(-2 \%/L)$[b]

[a]Flexibility (20 % elongation); [b]2 % reduction per layer

by hydrogen atoms. Ever since graphene, one of new emerging advanced carbonaceous materials, was discovered in 2004 by Geim and Novoselov (Novoselov et al. 2004) in University of Manchester, enormous research has been concentrated on this fascinating a single-layer 2D honeycomb carbon allotrope over the past few decades due to its exceptional physical, chemical, and mechanical properties igniting many exciting researches for various applications as shown in Table 1.2 (Novoselov et al. 2004; Zhang et al. 2005; Aleiner and Efetov 2006; Jannik et al. 2007; Schedin et al. 2007; Son 2010).

This atomic layered material has a large surface area (~ 2630 m²/g) (Zhou et al. 2010a, b, 2011a, b), high electron transport mobility (2×10^5 cm²/V s at RT) (Shen et al. 2012; Zhou et al. 2010a, b), larger mechanical strength (tensile strength: 130 GPa/Young's modulus: 1000 GPa) and high flexibility (20 % elongation) (Zhou et al. 2011a, b), high optical transparency (97.7 %) (Bae et al. 2010) in the visible wavelength, extremely high thermal conductivity (5300 W/mK)/negative thermal coefficient (Geim 2009; Stankovich et al. 2006), and chemical stability (Xie et al. 2011). On the other hand, very small on-off ratio due to a nearly zero band gap and unsaturable source-drain characteristic current in bulk graphene are hurdles to hardly be overcome for the application of high speed and high frequency electronic nano devices, next generation alternative to silicon (Schwierz 2010). In the real application process, intrinsic 2-dim (D) graphene has been additionally limited due to easy agglomeration and poor dispersion in common organic solvents. This drawback has been solved by converting 2D into 1D graphene nanoribbons (GNRs)

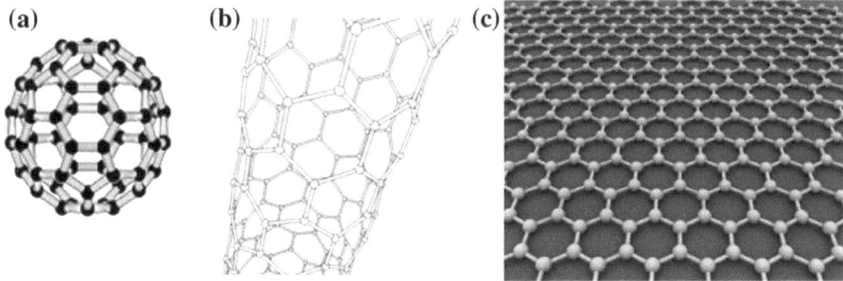

Fig. 1.6 **a** 0D C_{60} fullerene, **b** 1D SWCNT, and **c** 2D graphene

or into 0D graphene quantum dots (GQDs) by nanolithograhical tailoring and chemical synthetic approaches. Although GNRs show confined transport gaps and quantum dots associated with the geometry of the ribbons (Han et al. 2007; Todd et al. 2009; Stampfer et al. 2009; Molitor et al. 2009; Wang et al. 2011), recently much attentions have been paid on GQDs exhibiting unusual quantum confinement and edge effects. As represented in Fig. 1.6, C_{60}, CNT, and graphene can be classified as 0D, 1D, and 2D nanocarbons.

1.4   Metal Oxide-Nanocarbon Hybrid Materials

In energy-related electrochemical conversion and environment-related industry, precious noble metal-based materials have been widely used due to exceptionally high performance and stability. Alternative to these expensive and rare materials, lower cost and high performance metal oxides have been emerged as counterpart due to abundance, stability, and electrical and chemical capacities. In particular, enormous interests have been paid on metal oxide/nanocarbon hybrid materials in the electrochemical reactions on energy conversion and storage and their electrocatalyst systems. Moreover, these hybrid materials also showed prominently enhanced performance in supercapacitor electrode, Li ion secondary battery anode, photoanode in electrochemical reaction, photovoltaic solar cells and light emitting diodes. Previously a simple hybrid structure of CNT/metal oxide heterojunction such as SnO_2/CNT (Han 2003), RuO_2/CNT, (Park 2003), and CO_3O_4/CNT (Shan 2004) were reported and other many application are well reviewed (Hu and Guo 2007). As a functional hybrid material, ZnO/MW CNT were synthesized and evaluated in UV photovoltaic cells (Li et al. 2009) where ZnO nanoparticles and MW CNT served as UV absorption materials and charge transport layer respectively. Carbon-metal oxide composition for supercapacitor electrode. Over the past few years a great number of CNTs/metal oxide heterostructures, such as TiO_2/CNTs, Co_3O_4/CNTs, Au/CNTs, Au/TiO_2/CNTs, Co/CoO/Co_3O_4/CNTs (Fig. 1.7) etc., were synthesized and their electrochemical properties were investigated as well.

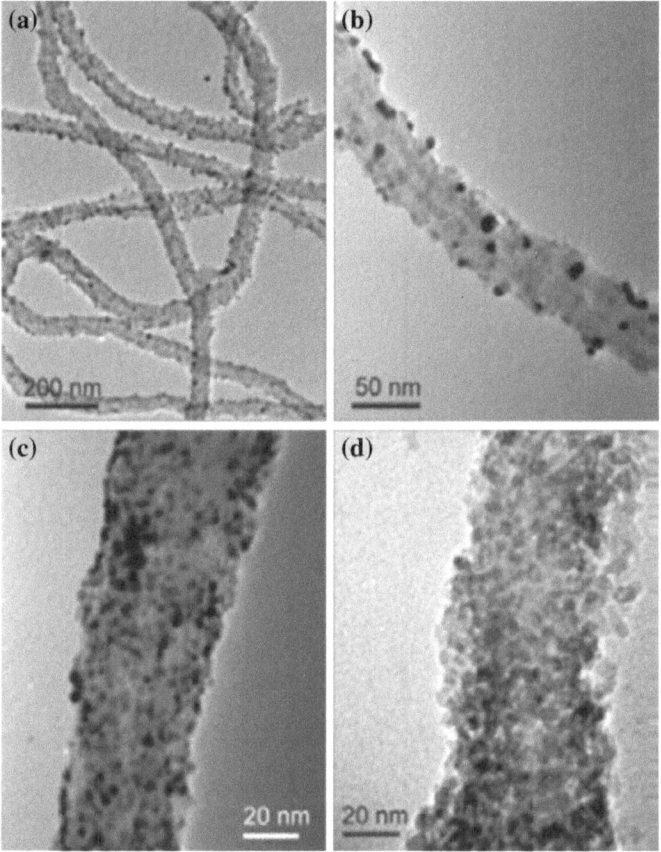

Fig. 1.7 **a**, **b** a large view and a detailed view on the Au/TiO$_2$/CNTs nanocomposites prepared by the photo-assisted method, **c** a detailed view on the Au/TiO$_2$/CNTs prepared by the self-assembly method, and **d** a detailed view on the TiO$_2$/Co$_3$O$_4$/CNTs nanocomposite prepared by the self-assembly method (Reproduced from Li et al. 2007)

Nanoparticle Au and MnO$_2$ for electrochemical supercapacitor, and MnO$_2$/In$_2$O$_3$ nanowires (NWs)/SWCNT for flexible asymmetric supercapacitor (ASCs) with advantages of mechanically flexible, mesoporous surface modifier and uniform surface morphology (Chen et al. 2010) Recently metal oxide hybrid materials with activated carbon (AC) as 0D, CNT and carbon fibers as 1D, graphene and reduced graphene oxide (rGO) as 2D, and porous carbon architecture as 3D for supercapacitor electrodes was well reviewed (Zhi et al. 2013). As shown in Fig. 1.8, RuO$_2$ anchored hybrid structure to graphene and CNT hybrid foam (RGM) was also synthesized for supercapacitor (Wang et al. 2014). By systematic approach to specific oxidation method, oxidation optimization, oxidation degree to balance inorganic-carbon coupling and the electrical properties were carried out to get strongly coupled metal oxide/nanocarbon hybrid. As application for electrocatalyst, Ni(OH)$_2$/CNT and

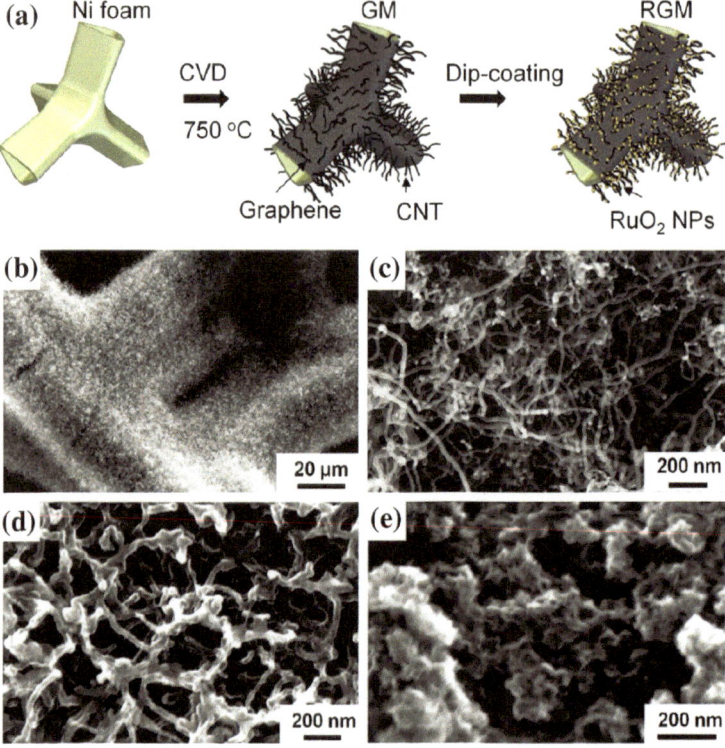

Fig. 1.8 a Schematics of the preparation process of RGM nanostructure foam. SEM images of **b**, **c** as-grown GM foam, **d** Lightly loaded RGM, and **e** heavily loaded RGM (Reproduced from Wang et al. 2014)

FeO_x/graphene hybrid materials was used for Ni-Fe battery, and CO_3O_4/rmGO (mildly oxidized graphene oxide), $MnCo_2O_4$/N-doped graphene, and MoS_2 or even CoS/rGO (reduced graphene oxide) chalcogenide hybrid were adopted for improving ORR (oxygen reduction reaction) (Linag et al. 2013). As a basic study to understand the mechanism of the formation of metal glycolate precipitate, mesoporous metal oxide-nanocarbon hybrid materials like TiO_2-, SnO_2-, Cu_2O/CuO-coated CNT, and TiO_2-coated graphene sheet was synthesized by a polyol-mediated self-assembly method and evaluated for Li-ion battery anode (Feng et al. 2014).

Another interesting application, nanostructured hybrids of carboneous materials such as amorphous carbon, CNT, and rGO with metal oxides (CoO, Fe_2O_3, V_2O_5) were prepared and evaluated as Li ion secondary battery anode materials (Yang et al. 2011; Shi et al. 2012).

Recently, hybrid structure of metal oxides like TiO_2 and ZnO on 2D graphene sheets were intensively investigated. TiO_2-graphene (TiO_2-GR) nanocomposite (Fig. 1.9a) showed much higher stability and activity than that of the bare TiO_2 for gas-phase photocatalytic degradation, where the performance of TiO_2-GR was

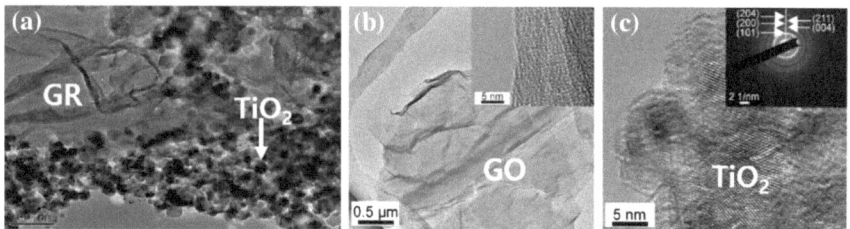

Fig. 1.9 **a** TEM images of TiO$_2$-GR (reproduced from Zhang et al. 2010), **b** graphene oxide (GO) and graphene oxide/TiO$_2$ (GOT) nanocomposite (Reproduced from Chen et al. 2010)

proved as similar to that of TiO$_2$-nanocarbons (CNT, C$_{60}$) (Zhang et al. 2010). Also nanocomposite of TiO$_2$ on graphene oxide (Fig. 1.9c) was synthesized and its photocatalytic performance was tested at the time of visible-light absorption.

Hollow ZnO nanospheres on reduced graphene oxide (rGO) nanocomposite as shown in Fig. 1.10d were synthesized for enhanced photocurrent response and photocatalytic activity. These nanocomposites showed better performance than hollow ZnO itself (Fig. 1.10a) which can be ascribed to the enhanced conductivity (Luo et al. 2012).

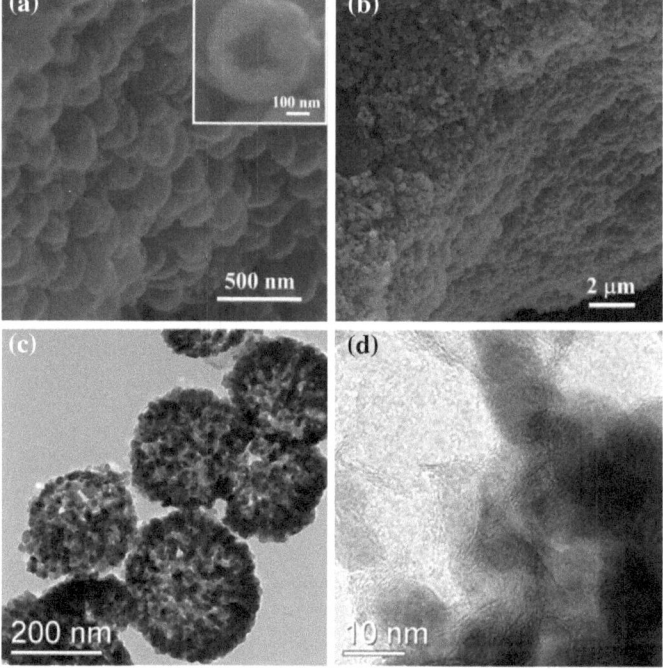

Fig. 1.10 Field emission scanning electron microscope images (**a**, **b**) FESEM images of hollow ZnO and TEM images (**c**, **d**) of RGO-ZnO 3.56 % (Reproduced from Luo et al. 2012)

As mentioned above, huge amounts of metal oxide/nanocarbon nanocomposite have been synthesized by various methods and evaluated by fabricating optoelectronic devices, electrochemical devices, and energy storage devices etc. Despite many efforts to manipulate more uniform metal oxide-nanocarbon nanocomposite structures, the metal oxide nano particles were still randomly scattered and nonuniformly attached to the nanocarbon surfaces. Until now, consolidated core–shell structure metal-oxide NPs encircled by nanocarbons with high conformality have been rarely reported.

References

I.L. Aleiner, K.B. Efetov, Phys. Rev. Lett. **97**, 236802–236804 (2006)
A.F.M. Anwar et al., IEEE T. Electron. Dev. **48**, 567 (2001)
S.K. Bae et al., Nat. Nanotechnol. **5**, 574 (2010)
J.R. Baena et al., Trends Anal. Chem. **21**, 187 (2002)
S. Baruah, J. Dutta, Sci. Technol. Adv. Mater. **10**, 013001 (2009)
J.W. Berry, A.J. Deutschman Jr., U.S. Patent 3,615,264A, 1971
C. Chen et al., ACS Nano. **4**, 6425 (2010)
J. Chen et al., J. Mater. Chem. **22**, 5164 (2012)
J. Chen et al., Sci. Rep. **4**, 4085 (2014)
K.S. Cho et al., Nat. Photonics **3**, 341 (2009)
S.M. Chou et al., Sensors **6**, 1420 (2006)
X. Dai et al., Nature **515**, 96 (2014)
M.S. Dresselhaus, *Science of Fullerenes and Carbon Nanotubes* (Academic Press, San Diego, 1995), pp. 112–115
B. Feng et al., Nanoscale **6**, 1437 (2014)
A.K. Geim, Science **324**, 1530 (2009)
S. Gullapalli, M.S. Wong, Chem. Engin. Prog. **107**, 28 (2011)
M. Guo et al., J. Solid State Chem. **178**, 1864 (2005)
W.Q. Han, Nano Lett. **3**, 681 (2003)
M. Han et al., Phys. Rev. Lett. **98**, 206805 (2007)
Y. He, Y. Li, Phys. Chem. Chem. Phys. **13**, 1970 (2010)
M.D. Ho et al., ACS App. Mater. Interface. **5**, 12369 (2013)
S. Hofle et al., Adv. Mater. **26**, 2750 (2014). doi:10.1002/dama.201304666
Y. Hu, C. Guo, in *Nanotechnology and Nanomaterial*, ed. by M. Naraghi (Intech, Rije ka, 2011), p. 3
J.M. Hvam, J. App. Phys. **49**, 3124 (1978)
C.M. Jannik et al., Nature **446**, 60 (2007)
D. Kabra et al., Adv. Mater. **22**, 3194 (2010)
R.G. Kepler et al., Appl. Phys. Lett. **66**, 3618 (1995)
J.Y. Kim et al., Adv. Mater. **18**, 572 (2006)
Y.H. Kim et al., Adv. Funct. Mater. **24**, 3808 (2014)
H.H. Kim et al., Sci. Rep. **5**, 8968 (2015a)
J.H. Kim et al., Nanoscale **7**, 5363 (2015b)
D. Klimm et al., Compr. Semicond. Sci. Technol. **3**, 302 (2011)
C. Klingshirn, ChemPhysChem **8**, 782 (2007)
D. Kohl et al., Surf. Sci. **41**(403), 403 (1974)
A. Kolmakov et al., Adv. Mater. **15**, 997 (2003)
H.W. Kroto et al., Nature **318**, 162 (1985)
J. Kwak et al., Nano Lett. **12**, 2362 (2012)

A.K.K. Kyaw et al., Adv. Mater. **25**, 2397 (2013)
G. Lalwani, B. Sitharaman, Nano LIFE **3**, 1342003 (2013)
J. Li et al., Nano Lett. **3**, 597 (2003)
J. Li et al., J. Am. Chem. Soc. **129**, 9401 (2007)
F. Li et al., Appl. Phys. Lett. **94**, 111906 (2009)
Y. Linag et al., J. Am. Chem. Soc. **135**, 2013 (2013)
B. Liu, C. Zeng, J. Am. Chem. Soc. **125**, 4430 (2003)
D.C. Look, Mater. Sci. Eng. B **80**, 383 (2001)
Q.-P. Luo et al., J. Phys. Chem. C **116**, 8111 (2012)
B.S. Mashford et al., Nat. Photonics **7**, 407 (2013)
F. Molitor et al., Phys. Rev. B **79**, 075426 (2009)
H. Morkoc et al., J. Appl. Phys. **76**, 1363 (1994)
K.S. Novoselov et al., Science **306**, 666 (2004)
J.-M. Ntep et al., J. Cryst. Growth **207**, 30 (1999)
E. Osawa, *Superaromaticity*, vol. 25 (Kyoto, Kagaku, 1970), pp. 854–863
Ü. Özgür et al., J. Appl. Phys. **98**, 041301 (2005)
J.H. Park, J. Electrochem. Soc. **150**, A864 (2003)
L. Qian et al., Nano Today **5**, 384 (2010)
L. Qian et al., Nat. Photonics **5**, 543 (2011)
R. Qiao et al., Nano Lett. **7**, 614 (2007)
F. Schedin et al., Nat. Mater. **6**, 652 (2007)
D. Schulz et al., J. Cryst. Growth **296**, 27 (2006)
F. Schwierz, Nat. Nanotechnol. **5**, 487 (2010)
Y. Shan, Chem. Lett. **33**, 1560 (2004)
J. Shen et al., J. Mater. Chem. **22**, 545 (2012)
W. Shi et al., J. Phys. Chem. C **116**, 26685 (2012)
F. Solis-Pomar et al., Nanoscale Red. Lett. **6**, 524 (2011)
D.I. Son, Nano Lett. **10**, 2441 (2010)
C. Stampfer et al., Phys. Rev. Lett. **102**, 056403 (2009)
S. Stankovich et al., Nature **442**, 282 (2006)
Z.N. Tan et al., Nano Lett. **7**, 3803 (2007)
G.P. Tegos, Chem. Bio. **12**(1127), 1127 (2005)
M.W. Thesen et al., J. Poly. Sci. **A48**, 3417 (2010)
K. Todd et al., Nano Lett. **9**, 416 (2009)
X. Wang et al., Nano Lett. **9**, 3137 (2009)
C. Wang et al., Sensors **10**, 2088 (2010)
M. Wang et al., ACS Nano **5**, 8769 (2011)
W. Wang et al., Sci. Rep. **4**, (2014)
X.J. Xie et al., J. Mater. Chem. **21**, 2057 (2011)
W. Xu et al., Opt. Lett. **39**, 426 (2014)
Z. Yang et al., J. Mater. Chem. **C21**, 11092 (2011)
T. Yao, S.-K. Hong, *Oxide and Nitride Semiconductors: Processing, Properties and Applications* (Springer, Berlin Heidelberg, 2009), p. 37
H.B. Zeng et al., Adv. Func. Mater. **20**, 561 (2010)
Y.B. Zhang et al., Nature **438**, 201 (2005)
K. Zhang et al., Nanotechnol. **19**, 215706 (2008)
Y. Zhang et al., ACS Nano **4**, 7303 (2010)
F. Zhang et al., Sol. Energy Mater. Sol. Cells **97**, 71 (2012)
M. Zhi et al., Nanoscale **5**, 72 (2013)
K. Zhou, Electochim. Acta. **55**, 3055 (2010a)
K. Zhou et al., New J. Chem. **34**, 2950 (2010b)
K. Zhou et al., Electroanalysis **23**, 862 (2011a)
K. Zhou et al., New J. Chem. **35**, 353 (2011b)
Y. Zhou et al., Science **336**, 327 (2012)

Chapter 2
ZnO–Nanocarbons Core–Shell Hybrid Quantum Dots

2.1 ZnO–Graphene Quasi Core–Shell Hybrid Quantum Dots

2.1.1 Synthesis

Surface of graphite powder (Alfa Aesar, 5 g) was modified by ultrasonically by using 120 ml of a 1:3 mixture of HNO_3 (17 M) and H_2SO_4 (18 M) for 2 h at a power level of 200 W, while maintaining the temperature at 45 °C and functionalized oxygen-contained moieties such as epoxy (C–O–C), carboxyl (–COOH), and hydroxyl (–OH) etc. The dispersion was maintained at room temperature for 4 days, during which time the color of the dispersion became charcoal. The dispersion was then washed repeatedly with water in a cycle of centrifugation and decantation, and finally with ethanol. The product was allowed to dry at 55 °C for 12 h. The final graphite power with oxygen-based functional groups had a greyish appearance and was not as shiny as the starting material (Son et al. 2012a, b, c).

Mixed acids H_2SO_4 and HNO_3-treated graphite oxide (GO) power (40 mg) was uniformly dispersed in 40 ml of DMF ultrasonically for 10 min. Zinc acetate dihydrate $[Zn(CH_3COO)_2 \cdot 2H_2O]$ (0.92 g) was dissolved in 200 ml of DMF, then the GO solution was added while continually stirring to form a stable precursor. Embryo ZnO quantum dots (QDs) were preferentially formulated in the early stage as a result of the chemical reaction through dehydration process as follows.

$$Zn(CH_3COO)_2 \times H_2O + (CH_3)_2NC(O)H$$
$$\rightarrow ZnO + (CH_3COO)_2CHN(CH_3)_2 + H_2O$$

And ZnO QDs uniformly grew up to the size of less than 10 nm in diameter as illustrated in Fig. 2.1.

W.K. Choi, *ZnO–Nanocarbon Core–Shell Type Hybrid Quantum Dots*,
Nanoscience and Nanotechnology, DOI 10.1007/978-981-10-0980-8_2

Fig. 2.1 **a** Scanning electron microscope image (×300,000) of **b** high resolution transmission electron microscope image of ZnO QDs

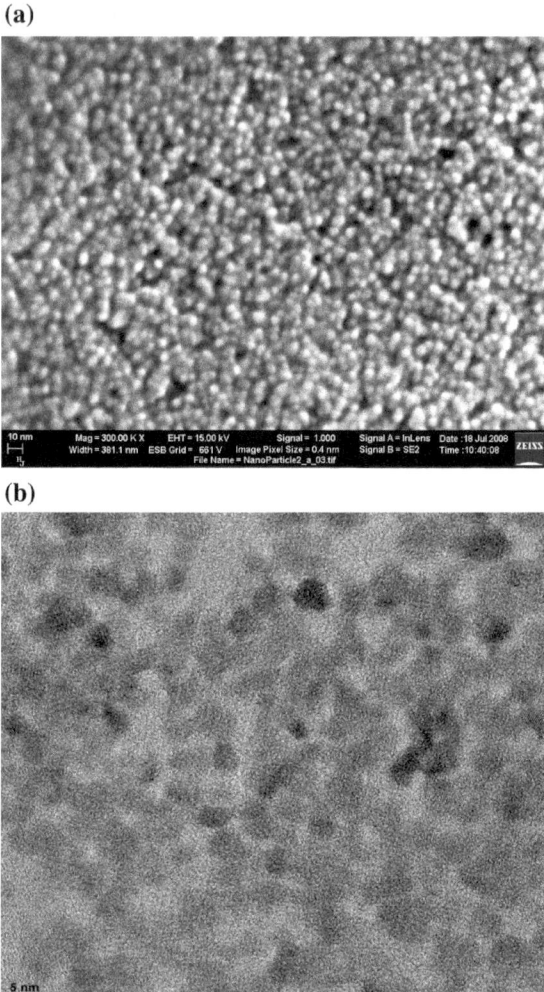

Subsequently, the mixed solution was heated to 95 °C and maintained at that temperature for 5 h. The color of the ZnO–graphene QDs then changed to white-greyish. This was subjected to repeated washing with ethanol by centrifugation, and finally with water. The final ZnO–graphene quantum-dot powder was obtained after drying the product at 55 °C as described in Fig. 2.2. In the formation of ZnO–graphene QDs, two chemical reaction are expected. One is chemical reaction between Zn^{2+} ions chemisorbed on the embryo ZnO QDs (denoted as $Zn^{2+}(ZnO)$) and the induced functional groups on the GO powder, leading to the local creation of new Zn–O–C bonding. Since ZnO is usually formed into oxygen-vacant nonstoichiometric oxide particles, $Zn^{2+}(ZnO)$ ions exist dominantly on the surface of ZnO QDs and easily react with oxygen groups induced on the GO

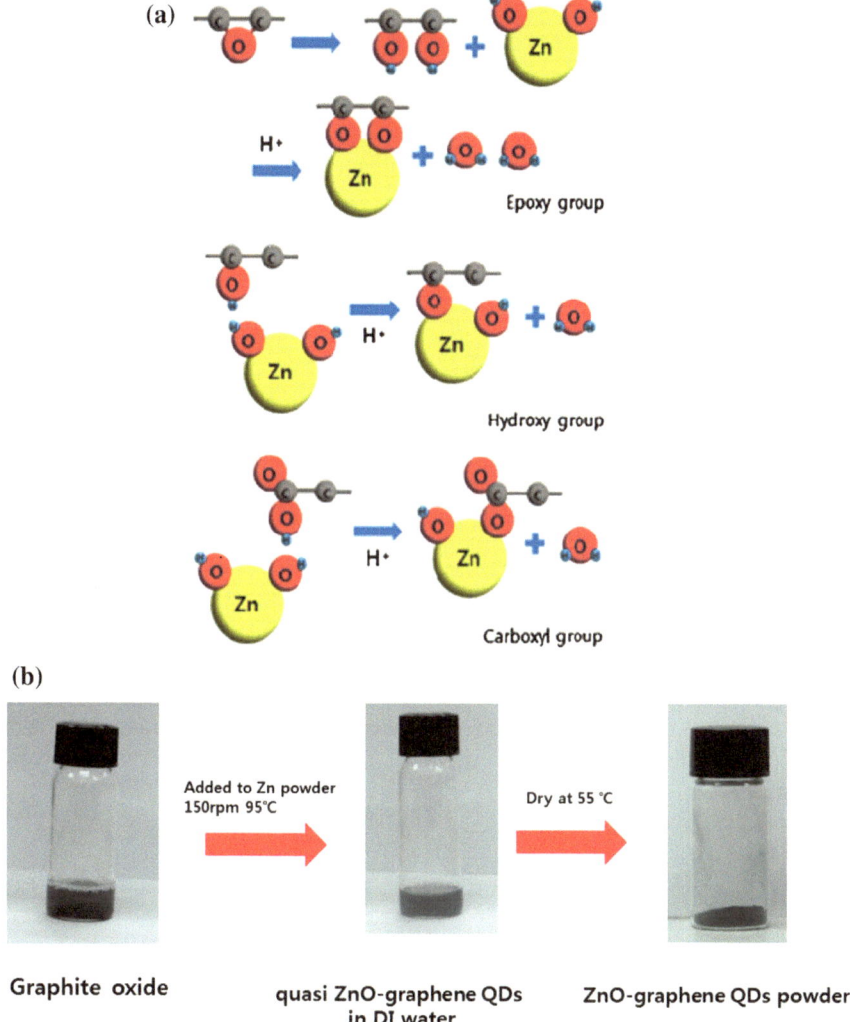

Fig. 2.2 a Chemical reaction model between $Zn^{2+}(ZnO)$ and oxygen-based functional groups on the graphite oxide (GO) powder. **b** Chemical process of the preparation of ZnO–graphene QDs

powder surface in the DMF solution, where all three different reaction kinds of functional groups, epoxy, hydroxyl, and carbonyl with $Zn^{2+}(ZnO)$ would produce the same Zn–O–C local bonding through similar dehydration process as shown in Fig. 2.3d.

In another reaction, Zn^{2+} ions bonded on the GO powder (denoted as $Zn^{2+}(GO)$) will react with O on ZnO QDs. During Zn–O–C bonding formation, the oxygen functionalities on the GO powder were almost reduced and a few layer graphene

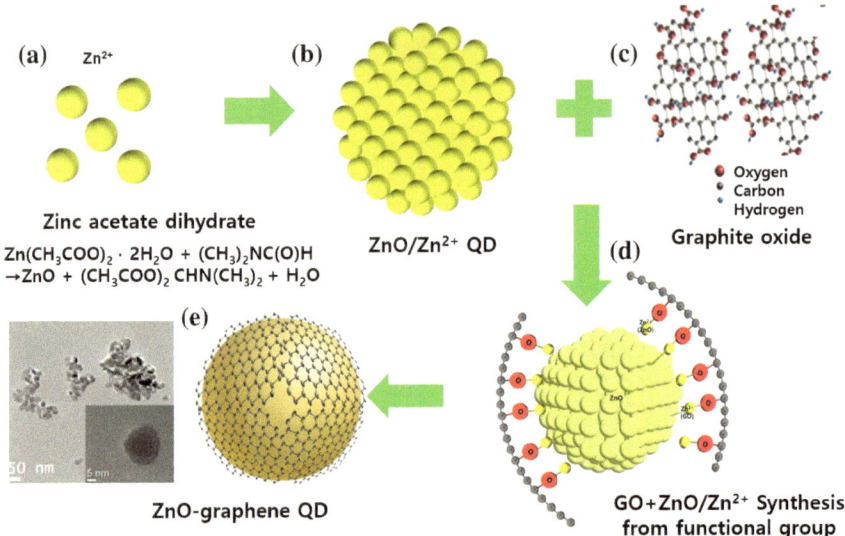

Fig. 2.3 Synthesis process for ZnO–graphene hybrid QDs. **a** Zn acetate dehydrate as a staring material to form ZnO NPs (**b**). **c** Chemical function group induced on graphite powder by acid treatment. **d** Synthesis of ZnO–graphene QDs through the formation Zn–O–C bonding between GO and Zn^{2+}. **e** Graphene QDs covered ZnO–graphene hybrid QDs and TEM image

quantum dots are detached from the GO powder via deoxidation, cutting process or a kind of chemical peel-off (i.e. chemical exfoliation) through the following reaction.

$$Zn(CH_3COO)_2 + 2H_2O \rightarrow Zn(OH)_2 + 2CH_3COOH$$
$$Zn(OH)_2 + Carbon\ functional\ groups\ (C-O-C,\ COOH,\ C=O) \rightarrow Zn-O-C + H_2O$$

In consequence, new functional nanoparticle of partially encircled ZnO–graphene consolidated hybrid quantum dots with quasi–core–shell structure can be synthesized. The whole synthetic model was conceptually depicted in Fig. 2.3. The chemical reaction of embryonic ZnO QDs with defect-like sites generated on GO powder surfaces plays a key role in stripping a monolayer or a few layer graphene quantum dot from the GO layers. The product of this reaction was then washed in water and dried to obtain pure ZnO–graphene quasi core–shell QDs.

2.1.2 Structural Characterization: XRD and TEM

Figure 2.4 shows the x-ray diffraction patterns for (a) pristine graphite powder, (b) GO powder treated without oxidant KMnO$_4$, and (c) that with KMnO$_4$ respectively. Main diffraction peak corresponding to G(002) from pristine graphite

Fig. 2.4 X-ray diffraction patterns for pristine graphite powder, GO powder treated by mixed acids H_2SO_4 and HNO_3 with/without oxidant $KMnO_4$

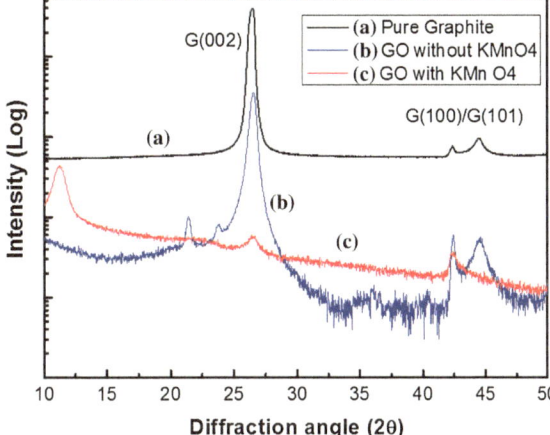

powder was strongly observed at $2\theta = 26.48°$ and the other diffraction peaks from G(100) and G(101) were also detected at $2\theta = 42.36°$ and $44.44°$. The GO powder treated without $KMnO_4$ showed similar XRD patterns to graphite powder without much change in diffraction peak intensity. This means that mixture of HNO_3 and H_2SO_4 without $KMnO_4$ less oxidized graphite powder and just moderately induced oxygen-based functional groups. When graphite powder was treated by $KMnO_4$ together, new diffraction peak around $2\theta = 11.26°$ was occurred and the peak corresponding to G(101) disappeared. The peak observed at the low angle near $2\theta = 10–12°$ is well known as fully oxidized graphite or graphene oxide (Son et al. 2012a, b, c). This revealed that $KMnO_4$ played an effective role for oxidizing graphite powder.

The HRTEM image of the ZnO–graphene quasi core–shell quantum dots is shown in Fig. 2.5a and clearly shows that the outer shell of the ZnO-core quantum dot looks like a single graphene layer. The enlarged view of the HRTEM image at the right bottom in Fig. 2.5b from the layer encircling by yellow color reveals that the interplanar spacing in the crystalline petal is 0.26 nm, which corresponds to the distance between two (002) planes of the hexagonal ZnO phase, indicating preferential growth along the [002] direction (c-axis). From another HRTEM analysis, close inspection of the dark areas in each of the images taken from the layer encircling the ZnO quantum dots by red color suggests that the layer is monolayer graphene, because a hexagonal atomic lattice with uniform contrast from the enlarged white color square region can be clearly discerned and the measured distance between carbon atoms of 0.14 nm agrees well with the theoretical value for graphene.

In Fig. 2.6, x-ray diffraction (XRD) patterns for the ZnO QDs, the ZnO–graphene quasi core–shell quantum dots, graphene sheets, and graphite powder are illustrated. For ZnO QDs, ZnO Bragg peaks (100), (002), (101) and (102) can be seen; the positions and intensity ratios of these peaks agree well with those of standard ZnO bulk (JCPDS no. 36-1451) and thus ZnO QDs were believed as being

Fig. 2.5 HR-TEM images for ZnO–graphene core–shell hybrid quantum dots. **a** Outer *black line* indicates graphene shells. **b** Enlarged picture of both white rectangular and *yellow circle* regions represents typical graphene hexagons and ZnO NPs

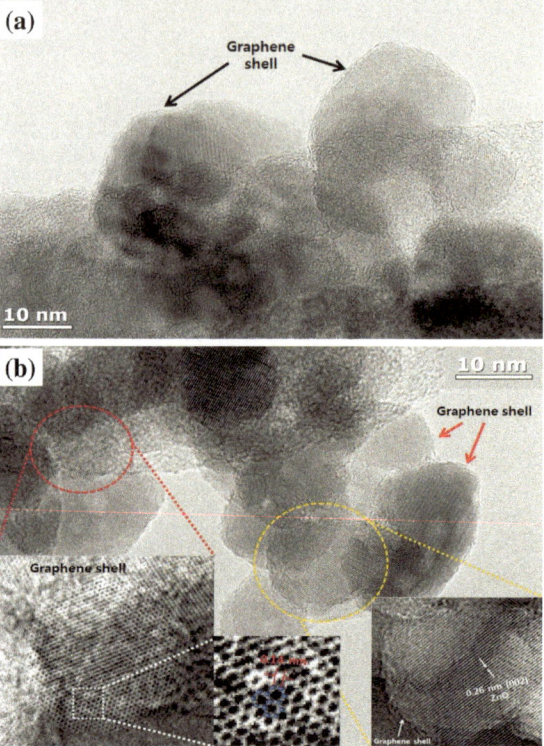

polycrystalline phase. For the ZnO–graphene quasi core–shell quantum dots, Bragg peaks corresponding to both ZnO and graphene are observed simultaneously, indicating that the consolidated ZnO–graphene quasi-core–shell quantum dots have been successfully synthesized from GO powder and zinc acetate dihydrate powder. The appearance of both a broad intense G (002) peak centered at $2\theta \approx 25.8°$ (d = 0.34 nm, with a large full-width at half-maximum (FWHM) as great as 2.34°) and a very broad peak denoted as G (100) at $2\theta = 43.5°$ with low intensity are strong evidence supporting for the existence of the graphene layer on the ZnO quantum dots. For comparison, XRD from graphite powder and graphene sheet are also added. XRD for graphite powder is the same with that aforementioned in Fig. 2.4. Graphene sheet is obtained by dissolving core ZnO QDs using HCl acid which will be discussed in Application section and is a kind of agglomerated graphene QDs chemically adsorbed on the inner ZnO. The diffraction peak form graphene sheet is happened at around $2\theta = 26°$ and shows relatively small intensity compared to that of graphite powder.

Figure 2.7a represents HRTEM images of ZnO–graphene QDs and enlarged images of white circle area (scanning line is depicted as white dot line in the right picture) Corresponding the energy dispersive x-ray spectroscopy (EDX) image of ZnO–graphene QDs was obtained by line scanning along the red line represented in

Fig. 2.6 X-ray diffraction patterns for the ZnO QDs, the ZnO–graphene quasi core–shell quantum dots, graphene sheets, and graphite powder

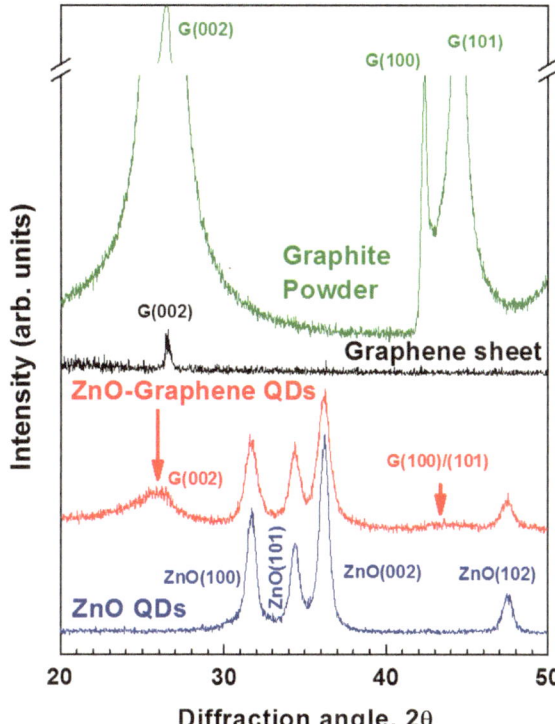

the left picture of Fig. 2.7b. In the right picture of Fig. 2.7b, peak intensity of C gradually increased at the edge but a little decreased at the center and reincreased after passing over the center. But those intensity of O and Zn showed the maximum at the center and symmetrically decreased at the edges. These results well support the existence of graphene layer outside ZnO core in ZnO–graphene QDs. From Thin Film Standardless Standard Quantitative Analysis (Fitting Coefficient: 0.6759) based on elemental x-ray emission peak of C K_α(0.277 keV) as reference, O K_α (0.525 keV), and Zn K_α (8.630 eV), atomic percentage was estimated as C (91.61 %):O(2.44 %):Zn(5.96 %), and corresponding compound as C(1):O(0.42): Zn(0.51) respectively (Fig. 2.7c).

In addition, a HR-HAADF (high resolution high angle annular dark field) STEM (scanning transmission electron spectroscopy) image (Fig. 2.8a) combined with an Energy Dispersive Spectrometry (EDS) image scanned across a single ZnO–graphene quasi core–shell QD were obtained. Figure 2.8b, d exhibit the drift corrected spectrum profile scanning EDS data of the C K line, O K line, and Zn L line of the ZnO–graphene quasi core–shell QDs, respectively. As seen in Fig. 2.8c, d, the intensities of both the O K and Zn L lines appear to show a sudden increase near 5 nm and the values are remain steady and then decrease after passing a position at near 20 nm. For the C K line spectrum (Fig. 2.8b), the intensity appears to increase after passing the position near 5 nm and then remains at a constant value.

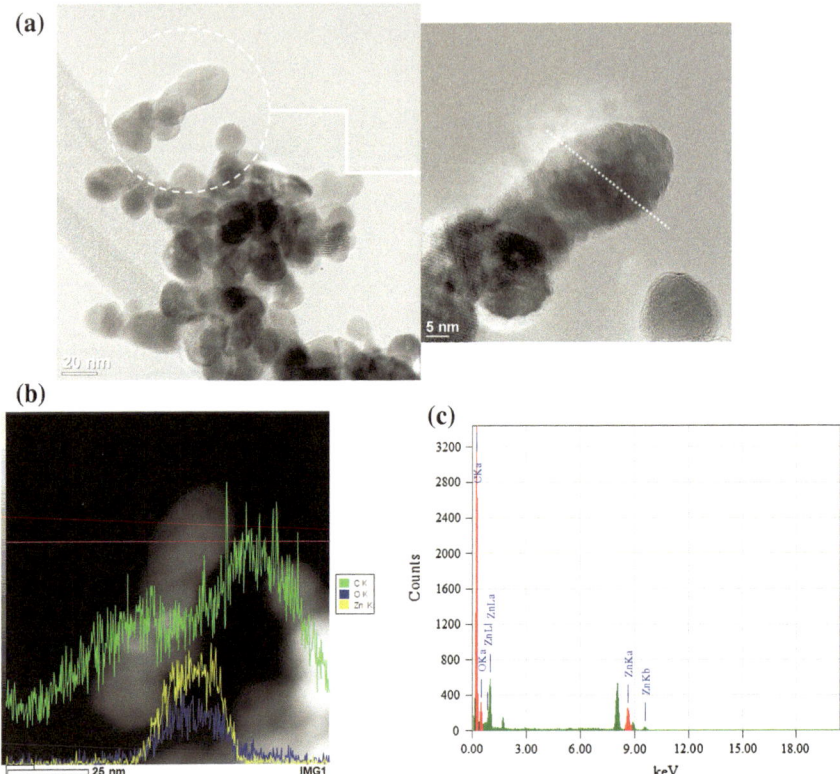

Fig. 2.7 TEM images of **a** ZnO–graphene core–shell hybrid QDs. **b** Scanned EDX profile corresponding to C K_α, O K_α, and Zn K_α along the red line and **c** EDX intensity

The increasing intensity, even after passing near 20 nm, might result from the overlap with the bottom edge of the particle. Both the HR-HAADF and scanned EDS data unambiguously show that the synthesized ZnO–graphene QDs are quasi consolidated structures of the ZnO QDs surrounded by a graphene layer.

2.1.3 Composition Analysis by X-Ray Photoelectron Spectroscopy

Figure 2.9 illustrates x-ray photoelectron spectroscopy data for mixed acids H_2SO_4 and HNO_3-treated graphite oxide (GO) power with and without strong oxidant $KMnO_4$ for comparison. By deconvolution of the C1s and O1s core level peaks into subpeaks, relative amounts of oxygen-related functional groups and the ratio of C/O were calculated (Son et al. 2012a, b, c).

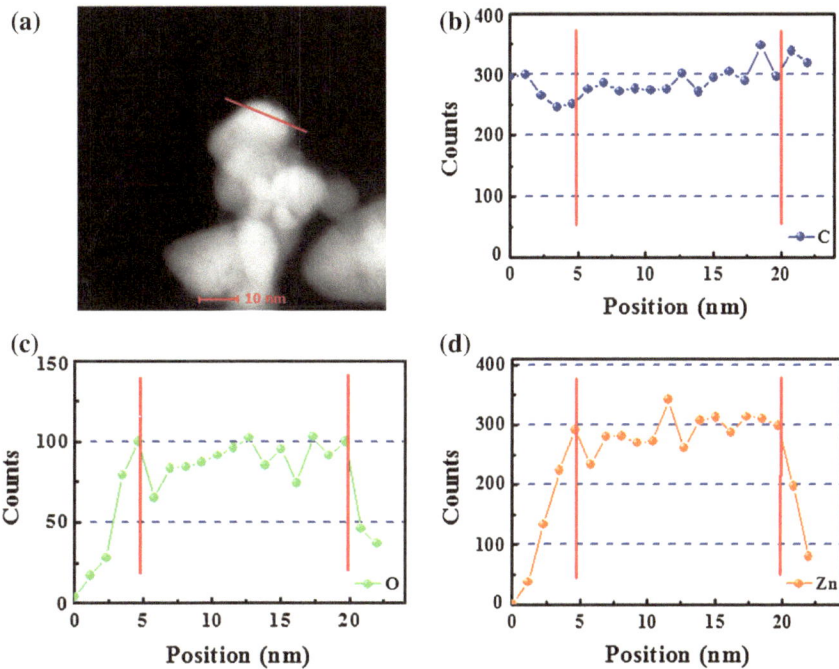

Fig. 2.8 HAADF STEM image **a** and spectrum profile scanning EDS of the C K_α line (**b**), O K_α line (**c**), and Zn L_α line (**d**) of ZnO–graphene quasi core–shell QDs (Reproduced from Son et al. 2012a, b, c)

Table 2.1 shows the amounts of carbon and oxygen, and the relative ratio C/O (XPS). The spectra were fitted after subtracting the Shirley background using the XPS Multipack Spectrum program. As presented in Table 2.1, the GO without addition of $KMnO_4$ showed the high C/O ratio of ~ 10, but that treated with $KMnO_4$ showed small as much as ~ 2 by large incorporation of oxygen contents. Even though the value of C/O of 10.07 is a little smaller than the value of 15.27 reported for RG-OHI-AcOH (Wang et al. 2008a, b) and 11.0 for highly reduced G–O (Moon et al. 2010), but was higher than the values of 9.97 (Cote et al. 2009) and 8.57 (Park et al. 2009) reported for chemically converted graphene.

On the other hand the value of 2.06 is close to the values of 1.4–1.7 reported for chemically exfoliated and highly oxidized graphene oxide nanoribbons obtained under various conditions using $KMnO_4$ (Li et al. 2008; Higginbotham et al. 2010), and similar to that of 2.5 for pristine graphene oxide obtained using similar chemical processes (Lomeda et al. 2008; Yang et al. 2009), and 2.1 using the modified Hummers method (Fan et al. 2011). The C1s signals were deconvoluted into sp2 C (284.5–284.6 eV), sp3 C (285.5–285.6 eV), C–O (286.1–286.7 eV), C=O (287.5–287.8 eV), –COO– (288.8–289.2 eV), and a π–π* satellite peak (290.6 eV) (Fan et al. 2011; Mattevi et al. 2009). From the XPS result, it can be

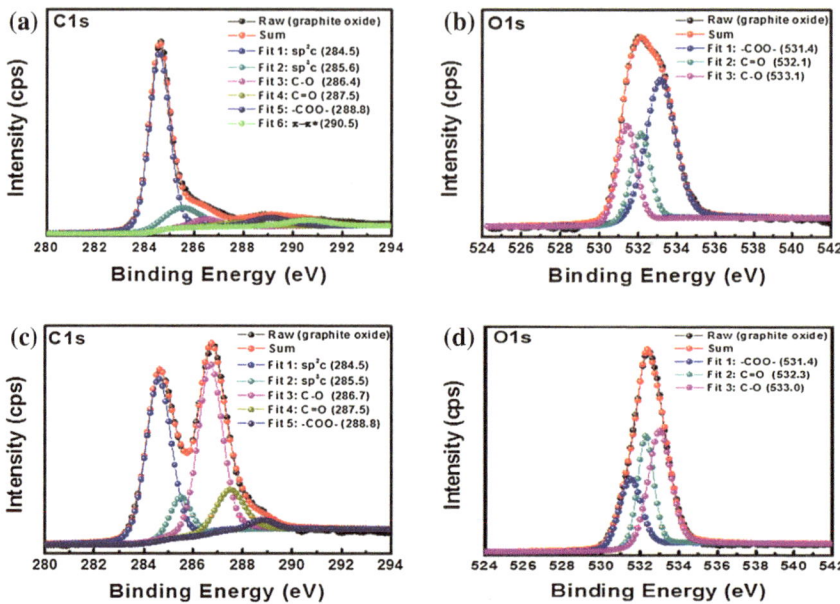

Fig. 2.9 XPS C1s (**a**) and O1s (**b**) core-level spectra for the GO treated with mixture H_2SO_4 and HNO_3 without $KMnO_4$ and C1s(**c**) and O1s (**d**) for the GO treated with mixture H_2SO_4 and HNO_3 and $KMnO_4$ (Reproduced from Son et al. 2012a, b, c)

Table 2.1 C and O contents and relative C/O ratio obtained from the XPS (Fig. 2.9) for the GO powder treated H_2SO_4 and HNO_3 with/without $KMnO_4$

Materials	C (%)	O (%)	C/O
Pristine graphite power	99.03	0.97	102
GO without $KMnO_4$	90.97	9.03	10.07
GO with $KMnO_4$	67.06	32.43	2.06

Reproduced from Son et al. (2012a, b, c)

concluded that mixed acids H_2SO_4 and HNO_3-treated graphite oxide (GO) power were not highly oxidized, but a certain intercalated graphite powder with sulfate or nitrate ions, similar to a graphite intercalation compound (GIC) (Tang et al. 2009; Nakajima et al. 1988).

2.1.4 Raman Spectroscopy

Figure 2.10a presents the Raman spectra of the ZnO–graphene quantum dots. The G and D modes are known to arise from first order scattering of the E_{2g} phonon of sp2 carbon atoms and from a breathing mode of k-point photons of A_{1g}

Fig. 2.10 Raman spectra for **a** ZnO–graphene core–shell hybrid QDs (Son et al. 2012a, b, c) and **b** single wall CNT. **c** Linear dependence of magnitude of bandgap opening of graphene on the strength of the applied uniaxial strain (Reproduced from Ni et al. 2008)

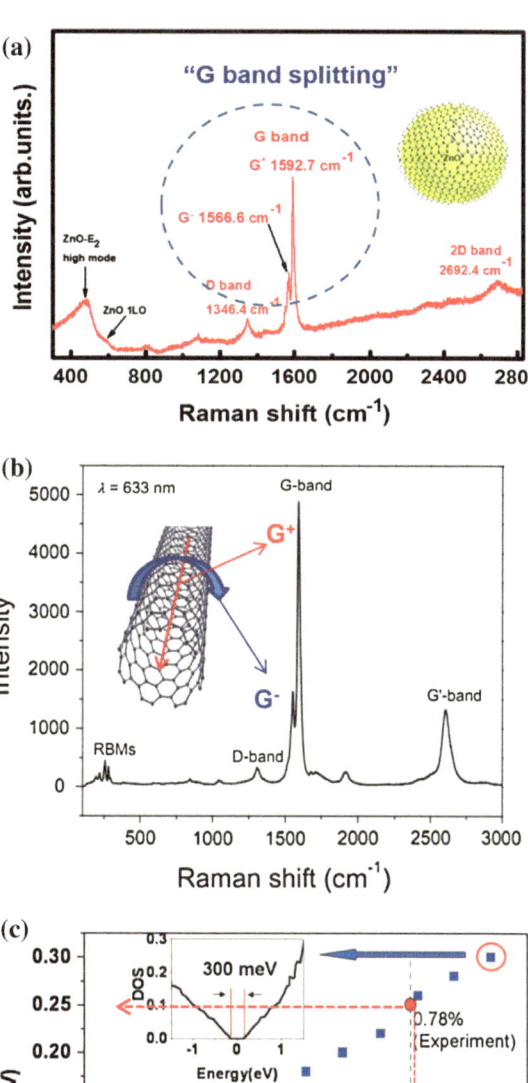

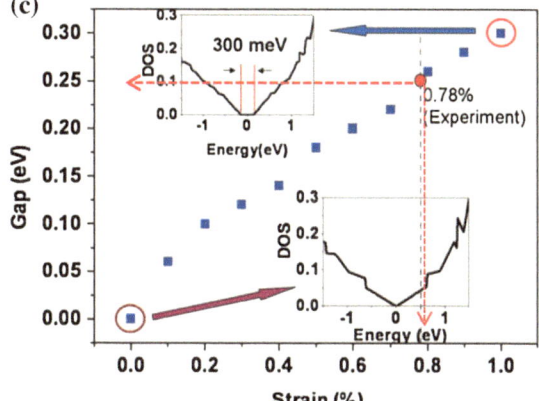

symmetry, respectively (Kudin et al. 2008). It is noteworthy that the doubly degenerate G peak splits into two sub-bands, namely G^+ (1592.7 cm^{-1}) and G2 (1566.6 cm^{-1}) (Mohiuddin et al. 2009), but this did not happen for the 2D peak. This splitting of the G band into two distinct sub-bands (G^+, G^-) results from strain induced symmetry breaking, with polarization along the strain as well as perpendicular to it, and the splitting increases with increasing strain. Such results are frequently observed in Raman spectra of single-walled CNT(SW CNT) as shown in Fig. 2.10b. The G band splitting becomes less pronounced as the CNT diameter increases and disappears for large CNT radii or for the case of multi-walled CNTs (Rodriguez et al. 2012). This phenomenon can be explained well in terms of strain induced by bending of the graphene surrounding the ZnO quantum dots. Even though, in reality, the strain is not uniaxial, the applied strain can be approximated from the splitting of 26.1 cm^{-1} as $\varepsilon = 0.8$ %, calculated from the best linear fit of the variations of the positions of G^+ and G^- as a function of applied uniaxial strain, $\partial\omega_{G+}/\partial\varepsilon \approx -10.8$ cm^{-1}/% and $\partial\omega_{G-}/\partial\varepsilon \approx -31.7$ cm^{-1}/%, provided uniaxial strain is applied. According to Ni et al. (2008), the magnitude of bandgap opening ($\Delta\varepsilon_g$) of graphene depends linearly on the strength of the applied uniaxial strain (as presented in Fig. 2.10c), and the induced strain of $\varepsilon = 0.8$ % applied here is equivalent to a bandgap opening of $\Delta\varepsilon_g \approx 250$ meV. Furthermore, the position of the sharp 2D peak at ~ 2692.4 cm^{-1} is further evidence that the graphene covering the ZnO quantum dots is a monolayer, because a single layer of graphene has a single sharp 2D peak below 2700 cm^{-1} (Dato et al. 2008).

2.1.5 Time-Resolved Photoluminescence

Figure 2.11 shows PL spectra at room temperature for the ZnO graphene quantum dots, poly-TPD and pure ZnO quantum dots respectively. The PL peaks for the pure ZnO quantum dots and poly-TPD are observed at wavelengths of 379 nm (3.26 eV) and 460 nm (2.69 eV), which well corresponds to the known values of excitonic emission with a bandgap (Anikeeva et al. 2003; Son et al. 2009). In the PL of the ZnO–graphene quantum dots, two additional emissions peaking at 406 nm (3.05 eV) and 432 nm (2.86 eV) can be observed, with the difference of peak positions apart from those of the pure ZnO quantum dots as much as ~ 190 meV. The conjugation of the graphene to the ZnO QDs leads to large extent of ca. 70 % of quenching of the UV PL emission with the appearance of new graphene peaks. The QD PL quenching effect may result from four different mechanism (Wang et al. 2008a, b). In this study, it can be presumed that the ZnO QD quenching in the presence of graphene is primarily caused by both the static quenching and charge transfer reactions. Unlike dynamic quenching, static quenching occurs when the donor and acceptor materials are in the ground state. Here, the inner ZnO QD is the donor and the graphene nanoshells are the acceptor. A possible quenching method is the fast deactivation of the excited state through the electron transfer reactions

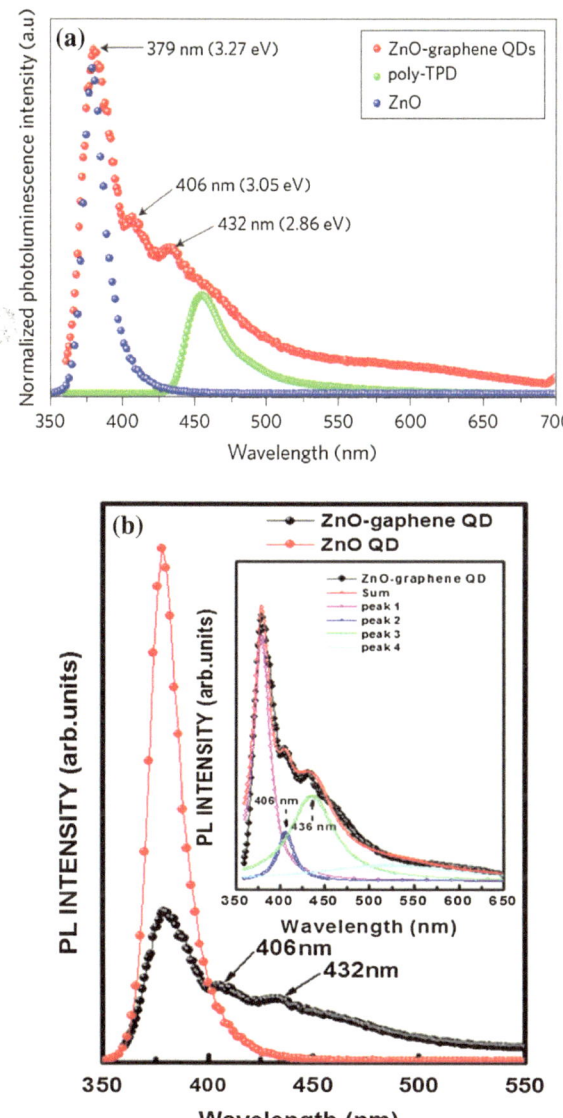

Fig. 2.11 **a** Photoluminescence spectra with normalized intensity for ZnO quantum dots, ploy-TPD, and ZnO–graphene quantum dots. **b** PL peak fitting with three subpeaks centered at 379, 406 and 436 nm (Reproduced from Son et al. 2012a, b, c)

from the conduction band of the ZnO to the lowest unoccupied molecular orbitals (LUMO) of the graphene quantum dots attached outside ZnO inner cores.

Qualitatively, this implies a high charge generation efficiency (Kumar et al. 2006). The three functional groups (carboxyl (–COOH), hydroxy (–OH), and

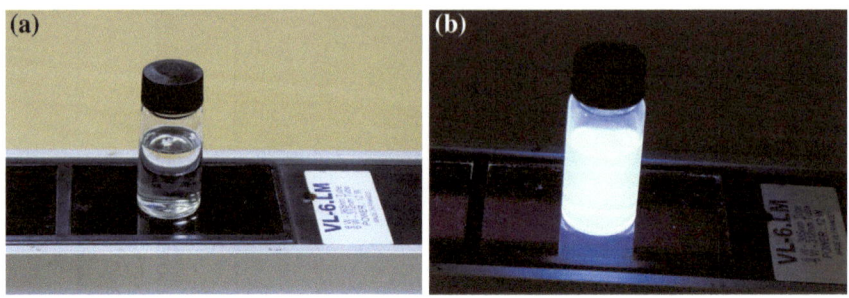

Fig. 2.12 Photographic images of transparent ZnO–graphene hybrid quantum dots solution dissolved in ethanol before (**a**) and after (**b**) irradiation of UV light (λ = 365 nm)

epoxy) formed on the GO surface can easily bind to the ZnO surface while forming the consolidated quasi core–shell QDs (which is confirmed in the subsequent analysis using XPS), resulting in strong PL quenching. Thus, the static quenching (strong quenching) of the QDs is caused by the three functional groups when it is attached to the ZnO surface and will be discussed further in the analysis of the time-resolved PL measurements.

Figure 2.12 is the photographic images of ZnO–graphene hybrid quantum dots dissolved in ethanol before (a) and after (b) exposure to UV light source (λ = 365 nm). As shown in Fig. 2.12a, ZnO–graphene quantum dots solution are transparent which means that they are well dispersed in ethanol without agglomeration. When this solution is exposed to UV light ((λ = 365 nm), very bright blue emission is observed as shown in Fig. 2.12b. This result strongly supports new blue emission observed in PL from ZnO–graphene hybrid quantum dots.

Fig. 2.13 The temporal evolution of the PL intensities taken by TRPL for the hybrid ZnO–graphene nanoshell QDs and for the ZnO QDs reference samples at 375, 383, 415, 460, and 500 nm (Reproduced from Son et al. 2012a, b, c)

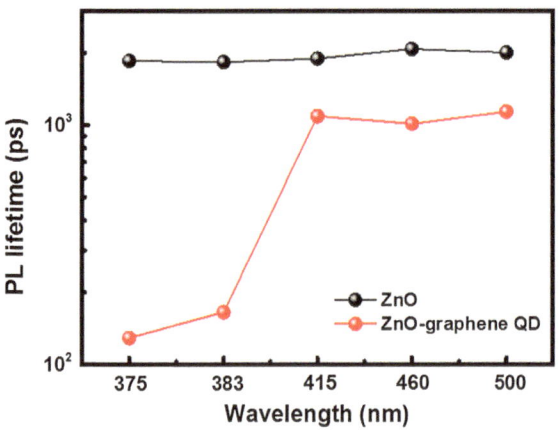

Table 2.2 Exciton lifetimes (τ_{avr}) of ZnO QDs and ZnO–graphene core–shell QDs

Samples	Monitored wavelength, nm	τ_1 (f_1), ns	τ_2 (f_2), ns	χ^2	τ_{avr}, ns
ZnO	375	3.56 (0.44)	0.53 (0.56)	1.34	1.85
ZnO	383	3.52 (0.44)	0.50 (0.56)	1.27	1.83
ZnO–graphene	375	1.25 (0.04)	0.08 (0.96)	1.073	0.13
ZnO–graphene	383	1.31 (0.07)	0.08 (0.93)	1.12	0.17

Reproduced from Son et al. (2012a, b, c)

Time-resolved PL measurements for both the ZnO–graphene quasi QD and the ZnO QD reference sample were performed in order to consider the charge transfer between the ZnO QDs and the graphene nanoshells. Figure 2.13 presents the temporal evolution of the PL intensities for the hybrid ZnO–graphene nanoshell QDs and for the ZnO QDs reference samples at 375, 383, 415, 460, and 500 nm, measured using a time-correlated single photon counting setup. The PL lifetimes of the nanocomposite decreased significantly in comparison with those of the reference ZnO QDs without conjugation to the graphene nanoshells. Table 2.2 shows the fitting curves of the PL decay with a biexponential function to calculate the lifetime in the UV range (375 and 383 nm). The amplitude weighted average exciton lifetime (τ_{avr}) was $f_1\tau_1 + f_2\tau_2$, where f_1 and f_2 are fractional intensities and τ_1 and τ_2 are lifetimes. χ^2 is the reduced chi-square. The lifetimes (τ) of the hybrid ZnO–graphene nanoshell QDs and the reference ZnO QDs at various wavelengths were fitted using a second-order equation and are plotted in Fig. 2.13. Considering that the decays are biexponential in nature, the amplitude-weighted average lifetimes were estimated. The calculated average lifetimes of the 375, 383, 415, 460, and 500 nm states for the ZnO QD and the ZnO–graphene QD were (1852, 129 ps), (1832, 165 ps), (1893, 1088 ps), (2077, 1012 ps), and (2004, 1136 ps), respectively. The decay of the hybrid ZnO–graphene nanoshell QDs at all selected wavelengths was much faster than that for the reference sample of ZnO QDs, and this is indicative of the existence of an additional high efficiency relaxation channel in the former, which is believed to be a charge transfer from the ZnO QD to graphene nanoshells.

2.1.6 Density of States (DOS) Calculated by Density Functional Theory (DFT)

A simple model with density functional theory (DFT) as implemented in the Gaussian package (Frisch et al. 2004) for calculation of density of states (DOS) and projected density of states (PDOS) was adopted to give details for the occurrence of the two new peaks in the PL spectra. The local structures of graphene were simulated with several oxygen bonding with 19 aromatic rings. The oxygen arrangement was achieved through several bond geometries, such as epoxy bonds and hydroxyl bonds in the basal plane and carboxyl groups at the edge, and the three cases were calculated. A Becke-style three-parameter Lee-Yang-Parr hybrid functional with a split basis set of 6-31G* was implemented to make calculations. A large periodic graphene model was not generated because this size of local model quite sufficiently to excerpt the core properties of the energy level changes associated with oxygen bonding geometries. First, a pristine structure with a planar geometry was generated to represent the graphene. Then, an oxygen atom with several bond types was attached as a model for the graphene–oxygen (G–O) bridging ZnO QD and the geometry fully relaxed to meet minimum energy requirements. Four different G–O models were selected and the DOS, PDOS, and molecular orbital characters were sequentially calculated for each G–O. The molecular structures of fully relaxed G–Os are depicted in Fig. 2.14a, e.

Figure 2.14f shows the calculated DOS of the graphene models. The magnified ($5\times$) PDOS of oxygen is indicated with the blue line and the Kohn–Sham molecular orbital energy is indicated by the vertical bars. Comparing the pristine graphene with the other G–Os, a perceptible change is observed in the LUMO region. The oxygen attachments induce significant orbital hybridization between the oxygen orbital and the p orbital of the pristine graphene; thus, the LUMO level of the pristine graphene splits (vertical arrows) into 2, 3 or 4 molecular orbitals. The oxygen PDOS clearly shows a contribution to the newly emerging levels; therefore, these levels are directly related to the oxygen attachment. These emergent energy levels could alter the orbital character of the pristine graphene. The PL process in the ZnO QD can also occur via another route: the oxygen atom connects the ZnO QD with the graphene and the photoexcited electrons from the valence band (O $2p$) of the ZnO QD are transferred to the unoccupied states of the G–Os. Then, these electrons undergo transition down to the original O $2p$ ground state leading to the PL emission. However, because the transition is satisfied by the selection rule ($\Delta l = \pm 1$), only electrons with s ($l = 0$) or d ($l = 2$) characteristics are allowed to transit to the O $2p$ ($l = 1$) level of the ZnO QDs. For pristine graphene, the LUMO level consists of the p orbitals of the aromatic rings; thus, the transition is prohibited during the PL process ($\Delta l = 0$). However, the newly emerged levels in the G–O having s orbital characteristics are occurred and the electrons in these levels can transit to the O $2p$ of ZnO QDs ($\Delta l = 1$). This corresponds well with the two newly emerged features in the PL spectra (406 and 432 nm) as shown in Fig. 2.11.

In order to quantify which oxygen bonding predominantly determines the PL transition, the percentage contributions of s and p orbitals to LUMO, LUMO+1, and LUMO+2 of the G–Os were evaluated. For Fig. 2.14b basal epoxy, LUMO and LUMO+2 have a definite contribution from s orbital (7.1 and 6.7 %, respectively; red arrows), while LUMO+1 (black arrow) has a negligible s orbital character. Meanwhile, for the 14(c) basal center, the LUMO has a significant s orbital character (19 %; red arrow) and its energetic position is identical to the LUMO of Fig. 2.14b basal epoxy. These levels have the three largest s characters among the levels in four different G–Os; thus, these three levels play a crucial role in the new features of the PL emission. Therefore, the photoexcited electron from the ZnO QD

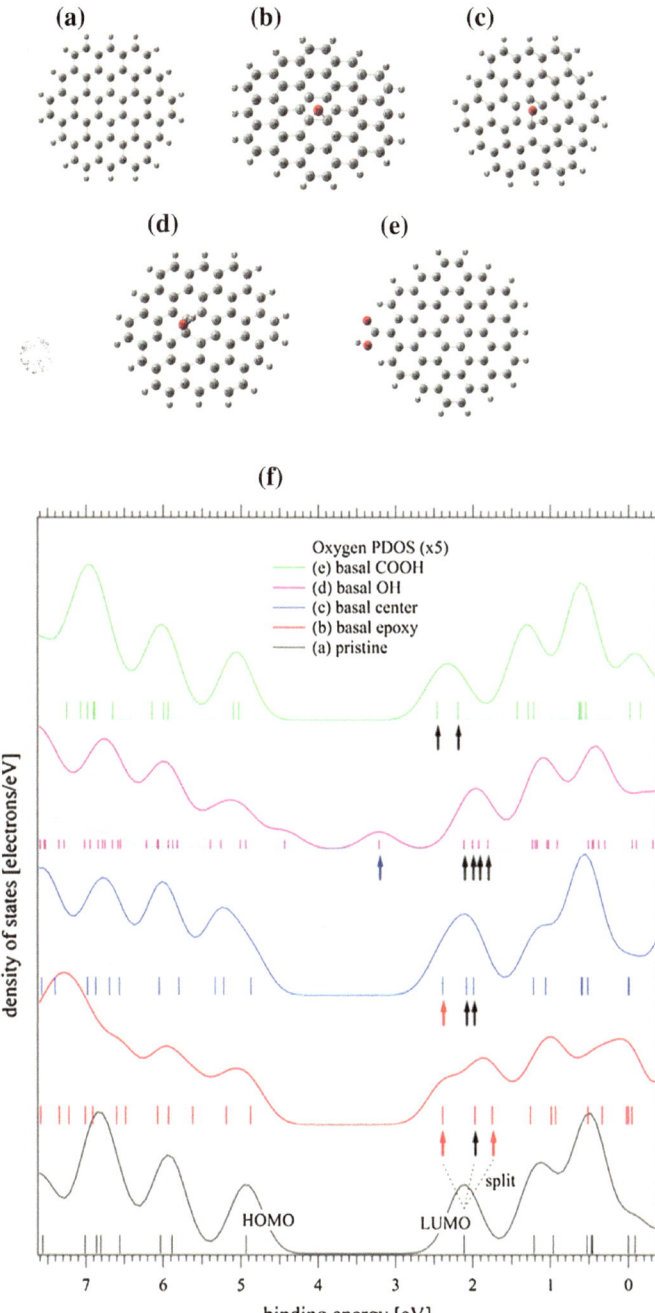

Fig. 2.14 Optimized structure of **a** pristine, **b** basal epoxy, **c** basal center, **d** basal –OH and **e** edge –COOH graphene model with 19 aromatic rings. **f** Calculated density of states (DOS) and oxygen projected DOS. LUMO splits are also depicted with *vertical arrows* (Reproduced from Son et al. 2012a, b, c)

can have a ground state transition using these new levels in the graphene shell, and it corresponds to the two new features in the PL spectrum, with the two levels being energetically identical.

Although the energetic distance between the newly emerged levels in the DFT results is larger than that of the PL emission, this does not significantly interfere with the underlying physics because the discrepancy is only a systematic error from the inherent properties of DFT, such as the energy gap underestimation. For Fig. 2.14d basal OH, only the LUMO (blue arrow) level has a definite s orbital contribution (5.8 %) and the other frontier levels have a negligible s character (<1 %). However, Fig. 2.14d basal OH is not found in the G–Os, as shown in the XPS core level spectra (Fig. 2.9). In addition, compared with the PL features, the energetic difference between the LUMO of Fig. 2.14d basal OH and the other levels of Fig. 2.14b basal epoxy and Fig. 2.14c basal center are too large. Therefore, Fig. 2.14d basal OH–graphene is not a dominant bridging structure of the ZnO–graphene core–shell. Interestingly, Fig. 2.14e edge COOH splits the pristine LUMO into two levels without s hybridization occurring. Therefore, the LUMO and LUMO+1 of Fig. 2.14e edge COOH do not contribute to the PL emission. Collectively, the new PL features originate primarily from the LUMO and LUMO +2 of Fig. 2.14b basal epoxy and the LUMO of Fig. 2.14c basal center, which bridge the ZnO QD to graphene (Son et al. 2012a, b, c).

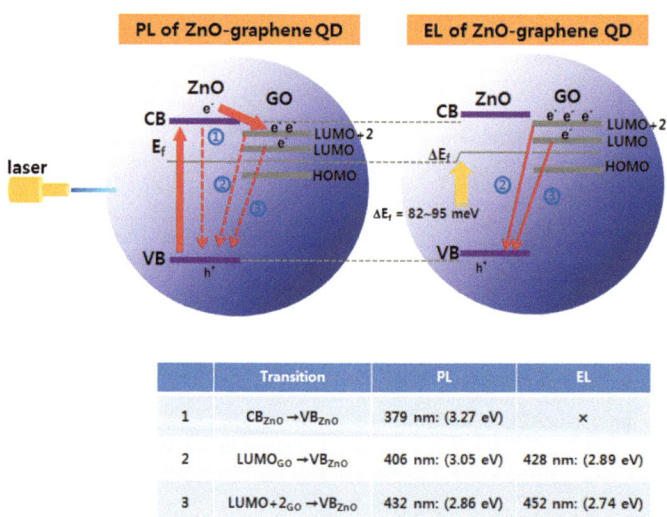

	Transition	PL	EL
1	$CB_{ZnO} \rightarrow VB_{ZnO}$	379 nm: (3.27 eV)	×
2	$LUMO_{GO} \rightarrow VB_{ZnO}$	406 nm: (3.05 eV)	428 nm: (2.89 eV)
3	$LUMO+2_{GO} \rightarrow VB_{ZnO}$	432 nm: (2.86 eV)	452 nm: (2.74 eV)

Fig. 2.15 Photoluminescence and electroluminescence transition scheme for ZnO–graphene quasi-quantum dots. Transitions 1, 2 and 3 correspond to electron transitions from the conduction band (CB) of ZnO, LUMO+2 and LUMO levels induced by G–O_{epoxy} to the valence band (VB) of ZnO, respectively (Reproduced from Son et al. 2012a, b, c)

Table 2.3 Percentage contribution of s and p orbitals to LUMO, LUMO+1 and LUMO+2 of four different kinds of graphene–oxygen (G–O) bonds

Ligand		s orbital contribution (%)	p orbital contribution (%)
Epoxy (–O)	LUMO (2.39 eV)	7.1	92.4
	LUMO+1 (1.97 eV)	0.6	99.1
	LUMO+2 (1.75 eV)	6.7	92.8
Centered epoxy (–O)	LUMO (2.39 eV)	19.4	80.2
	LUMO+1 (2.08 eV)	2.0	97.6
	LUMO+2 (1.99 eV)	2.3	97.2
Hydroxy (–OH)	LUMO(β) (3.21 eV)	5.8	93.7
	LUMO(α) (2.11 eV)	0.2	99.5
	LUMO+1 (β) (2.00 eV)	0.2	99.4
	LUMO+1 (α) (1.92 eV)	2.2	97.4
	LUMO+2 (β) (1.81 eV)	2.7	96.9
Carboxyl (–COOH)	LUMO (2.46 eV)	0.0	100.0
	LUMO+1 (2.19 eV)	0.0	100.0

Reproduced from Son et al. (2012a, b, c)

Figure 2.15 illustrates the schematic PL process from ZnO–graphene hybrid QDs electronic transition and EL process occurred in ZnO–graphene QDs based device. Electronic transitions from (LUMO+2)G–O$_{epoxy}$ or (LUMO)G–O$_{epoxy}$ to VB$_{ZnO}$ correspond to two PL peaks of 406 nm (3.05 eV) and 436 nm (2.84 eV) respectively and also correlate with two EL of 428 nm (2.89 eV) and 452 nm (2.74 eV) respectively (Table 2.3).

2.2 ZnO–C$_{60}$ Hybrid Quantum Dots

2.2.1 Synthesis

In a similar way as described in Sect. 2.1.1, C$_{60}$ oxide was prepared by treating C$_{60}$ natural powder with a mixed acid of H$_2$SO$_4$ and HNO$_3$ was treated and; thus, three functional groups, carboxyl (–COOH), hydroxy (–OH), and epoxy, could be induced on the C$_{60}$ surface. Thereafter, in the mixture of Zn acetate dihydrate with C$_{60}$ oxide in DMF solution, embryo ZnO QDs were preferentially formed in the early stages under the chemical reaction Zn(CH$_3$COO)$_2$·H$_2$O + (CH$_3$)$_2$NC(O)H → ZnO + (CH$_3$COO)$_2$·CHN(CH$_3$)$_2$ + H$_2$O and grew up to the size of about 5 nm in diameter. Then, at the outmost edge of the embryo ZnO QDs, two chemical reactions most likely occurred as shown in Fig. 2.16b. In one reaction, Zn^{2+} ions, denoted as Zn^{2+}(ZnO), chemisorbed on embryo ZnO QDs, react with functional groups on C$_{60}$ and as a result, creates local Zn–O–carbon bonding. In another

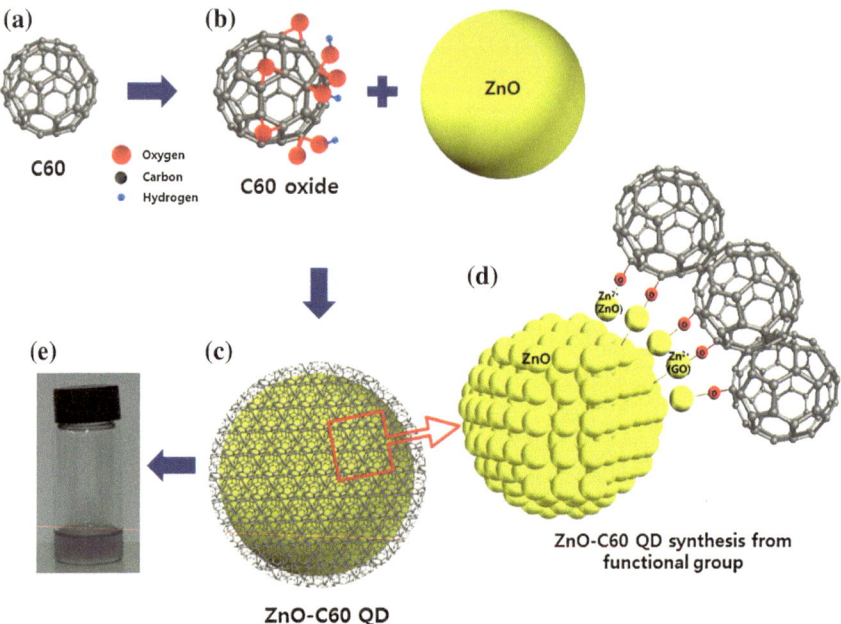

Fig. 2.16 The schematics of conceptual chemical synthesis of ZnO–C_{60} core–shell QDs. **a** Induced functional groups on the C_{60} oxide surface after treatment with an acid of H_2SO_4: $HNO_3 = 3:1$. **b** Embryo ZnO QDs formed under the chemical reaction and formation of the ZnO–C_{60} core–shell QDs. **c** Schematics of the ZnO–C_{60} core–shell QDs. **d** The possible chemical reactions between three kinds of functional groups (carboxyl (–COOH), hydroxy (–OH), and epoxy) and C_{60} oxide. **e** Image of the ZnO–C_{60} QDs in DMF solution (Reproduced from Son et al. 2012a, b, c)

reaction, Zn^{2+} ions ($Zn^{2+}(C_{60})$) bonded onto C_{60} also form Zn–O bonding and then combined with embryo ZnO QDs. Subsequently, the mixed solution was heated to 100 °C and maintained for 5 h in a constant temperature water bath and then cooled to room temperature. Finally, the ZnO–C_{60} core–shell QDs final product was obtained by using a filter.

2.2.2 Structural Analysis: X-Ray Diffraction and Transmission Electron Microscopy

Figure 2.17a shows a high-resolution transmission electron microscopy (HRTEM) image of an individual ZnO–C_{60} core–shell QD. The distorted hexagonal shaped area in the HRTEM image is actually showing C_{60} molecules on ZnO QDs with dimension of a long axis of about 17 nm and short axis of about 12 nm. The Zn–O–C bonding was possibly achieved through the defect like sites created on the acid treated C_{60} oxide and thus the chemical conjugation of C_{60} molecules with

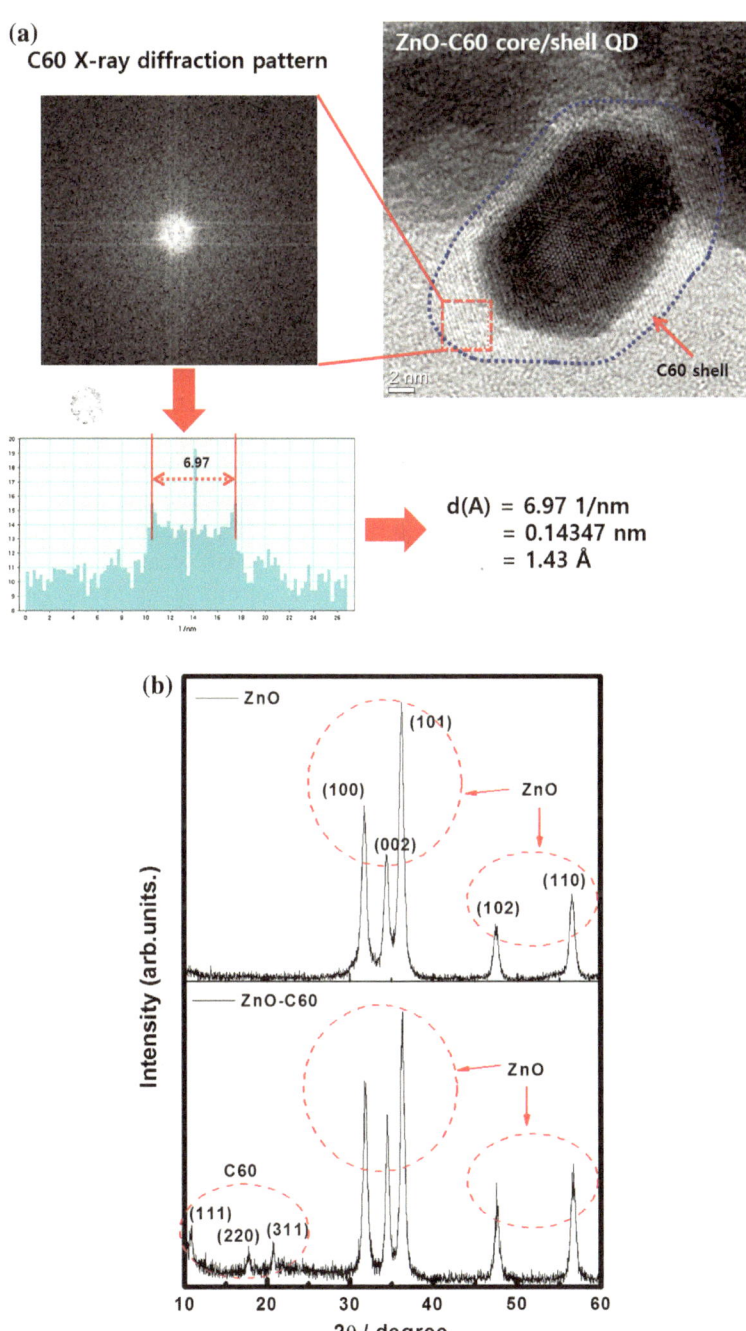

◀ **Fig. 2.17** HRTEM and X-ray diffraction of the synthesized ZnO–C$_{60}$ core–shell QDs. **a** The HRTEM image of the ZnO–C$_{60}$ core–shell QDs. The *red square line* is the image of one ZnO QD covered by C$_{60}$. The inset shows an enlarged view of C$_{60}$ on the surface of a single ZnO QD. The C–C bond length, although not clearly resolved everywhere, is measured to be 0.14 nm. **b** The appearance of ZnO and C$_{60}$ peaks in XRD patterns gives evidence that the ZnO–C$_{60}$ core–shell QDs include ZnO QDs and C$_{60}$

ZnO QDs. Further, from HRTEM analysis, the close inspection of the red squared area in the image reveals that the outer shell, encircling the ZnO QDs, is actually composed of C$_{60}$ because a hexagonal atomic lattice with uniform contrast can be clearly observed, and the measured distance between carbon atoms of about 0.14 nm coincides with the theoretical value of C$_{60}$. As seen in XRD patterns of ZnO–C$_{60}$ core–shell QDs (Fig. 2.17b), simultaneous observation of Bragg peaks of ZnO as well as C$_{60}$ indicates that the consolidated ZnO–C$_{60}$ core–shell QDs were successfully synthesized from C$_{60}$ oxide and zinc acetate dihydrate. The existence of C$_{60}$ on ZnO QDs is confirmed by the appearance of low intensity peaks of (111), (220), and (311) centered at around $2\theta = 10.9°$, $17.74°$, and $20.76°$, respectively and is in good agreement with the JCPDS (no. 44-0058) data. The ZnO (100), (002), (101), and (102) peaks, without any peaks corresponding to the Zn metal, were also clearly discernible and the position and intensity ratios of these peaks agree well with those of the ZnO bulk (JCPDS no. 36-1451).

2.2.3 Photoluminescence

The (PL) spectrum of ZnO QDs only shows a near-band-edge excitonic emission without a defect-related green band, as shown in Fig. 2.17. The peak at 381 nm (3.25 eV) is matching well with the emission peak observed for the ZnO bulk material (3.25 eV) (Son et al. 2009). On the other hand, the PL peak of ZnO–C$_{60}$ core–shell QDs is greatly reduced (quenched) as indicated by the black line in Fig. 2.17. This result is caused by the conjugation of C$_{60}$ with the ZnO QDs, leads to a PL quenching of about 99.8 % or almost no luminescence. This value is quite larger than ca. 70 % measured in the ZnO–graphene core–shell QDs. In similar way, this can be attributed to an efficient charge separation by direct electron transfer from the ZnO QDs to the C$_{60}$ through a chemical bonding of Zn–O–C. These results confirm the enhanced photo-induced charge transfer in ZnO–C$_{60}$ core–shell QDs. The inset in Fig. 2.18 represents the enlarged PL peak of ZnO–C$_{60}$ core–shell QDs. On the other hand, a new PL peak, not well resolved and not high intensity, can be clearly observed at around 400 nm besides the peak corresponding to ZnO. This can be due to recombination between the transferred electrons on the conduction band (4.3 eV) of C$_{60}$ and holes in the valence band (7.39 eV) of ZnO. The energy of this transition corresponds to about 3.09 eV and is coinciding with the peak observed. It is also noteworthy that, in addition to the large quenching of PL, the intensity of band-to-band transition corresponding to ZnO is lower than that

Fig. 2.18 PL spectra of ZnO only QDs (*red line*) and ZnO–C$_{60}$ core–shell QDs (*black line*). The inset shows the magnified PL spectrum of ZnO–C$_{60}$ core–shell QDs (Reproduced from Son et al. 2012a, b, c)

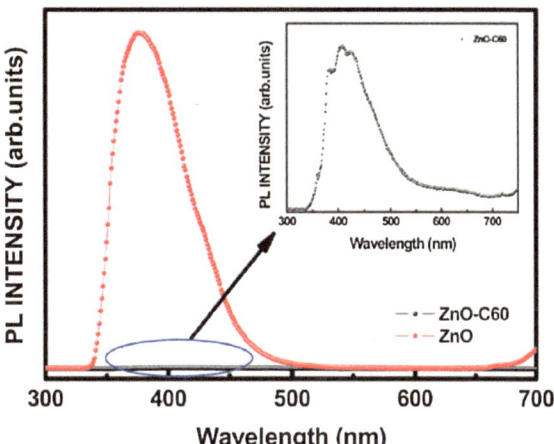

of transition between C$_{60}$ and ZnO, which supports that excited electrons are effectively transferred from ZnO to C$_{60}$. The QD PL quenching effect may occur by four mechanisms as described in Sect. 2.1.5 (Wang et al. 2008a, b). (1) dynamic quenching, (2) static quenching, (3) quenching by energy transfer, and (4) charge-transfer reactions (Bifeng et al. 2008). In the same way, ZnO QD PL quenching in the presence of C$_{60}$ is proposed to mainly be caused by static quenching and charge-transfer reactions. The other possible way of quenching can be due to the fast deactivation of the excited state by the electron transfer reactions from ZnO to C$_{60}$. The degree of quenching mainly depends on the number of electrons transferred from the conduction band of ZnO to LUMO levels of C$_{60}$ as well as the duration time on LUMO levels. Blue emission from LUMO levels of C$_{60}$ to the valence band of ZnO mainly consisted of O2p at 3.05 and 2.94 eV (see the inset in Fig. 2.11a, b. Almost complete quenching of all UV and blue emissions in PL indicates that most excited electrons in the conduction band of ZnO were transferred to cathode through high mobility C$_{60}$ before emission from LUMO in C$_{60}$ to the valence band of ZnO occurs. This qualitatively implies a high charge generation efficiency (Beek et al. 2004; Palthi et al. 2010).

2.3 ZnO–Nano Ring Single-Walled CNTs Hybrid Quantum Dots

2.3.1 Synthesis

The pristine single-walled carbon nano tubes (SWCNTs) synthesized by catalyst chemical vapor deposition (CCVD) are purity > 95 wt%, I$_G$/I$_D$ > 9 (Raman data), diameter(1–2 nm), and length ca. ~10 μm (Carbon Nano-Material Technology, Co.) (Fig. 2.19) SWCNTs were chemically functionalized with a mixture of

(a)   **(b)**

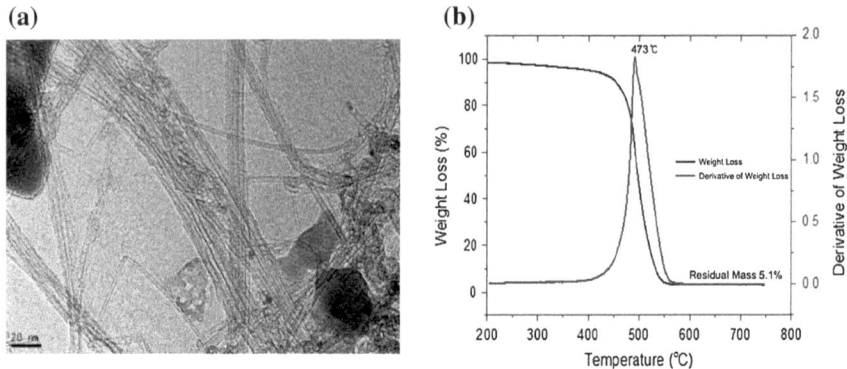

Fig. 2.19 High resolution transmission electron microscope (HRTEM) image (**a**) and TGA data (**b**) of CNTs SP95 (Carbon Nano-Material Technology Co.)

sulfuric and nitric acids (H_2SO_4:HNO_3 = 3:1). And then the SWCNT was reported to be cut into short tubes with most of carboxylic groups (–COOH) covalently attached the sidewalls, defects site, and open ends as illustrated in Fig. 2.20 (Banerjee et al. 2005). The SWCNTs were dispersed in acid mixture by using a multi frequency ultrasonicator. After acid treatment the solution was quenched in cold water, and samples were diluted. Thereafter, the SWCNTs were extracted on a PTFE membrane of 1um pore size using a vacuum filtration assembly. The resulting SWCNTs were dried in an oven at 70 °C for overnight.

In order to synthesize ZnO-NRCNTs nanoparticles, the solution of zinc acetate dehydrate (1.84 g) was prepared in 300 ml dimethylformamide (DMF) under stirring. Acid treated SWCNTs (20 mg) (Fig. 2.21a) were dispersed in 50 ml DMF by using a multi frequency ultrasonicator. The SWCNTs solution was added in the ZnO solution under continuous stirring for 5 h at 150 °C (Fig. 2.21b). The solution was poured into ethanol. After the reaction was quenched, the resulting gray solid products were washed with ethanol and water, centrifuged and then dried in the

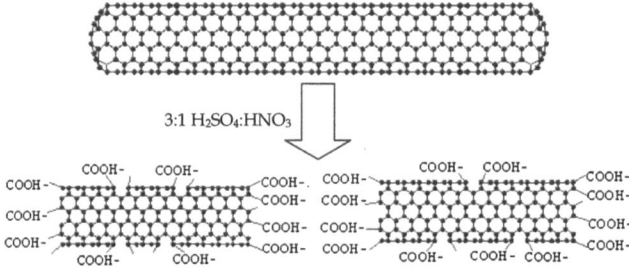

Fig. 2.20 A schematics of the chemical functionalization of a SWCNTs. The SWCNTs are cut into short tubes with carboxylic groups covalently attached the sidewalls and open ends (Reproduced from Banerjee et al. 2005)

(a) (b) (c)

Fig. 2.21 Synthetic process of ZnO–SWCNTs with PVP. **a** Neat CNTs were treated by acid. **b** Acid-treated CNTs was mixed with Zn acetate dehydrate in DMF solution. **c** ZnO–SWCNTs were reacted with PVP for 12 h

vacuum oven for overnight. For ZnO–NRCNTs blend with polyvinylpyrrolidone (PVP) were dissolved in water. In order to NRCNTs binding with PVP, each reagent was blended in solution by using a multi frequency ultrasonicator (Fig. 2.21c). After reaction, for removing unreacted polymer, the solution diluted in water and centrifuged. Resulting product dissolved in hydrochloride acid (HCl) and sonication. After reaction the solution poured into cold ethanol. After the reaction, the NRCNTs were washed with ethanol and water, centrifuged and dried in the vacuum oven for overnight.

2.3.2 Nano-ring SWCNTs (NR-SWCNTs)

Figure 2.22a shows high resolution TEM image of the ZnO-SWCNTs structure. The ZnO inner core is located inside and wrapped by the single-walled CNTs. ZnO is well crystallized as a hexagonal structure with the crystalline spacing of 0.26 nm along with the [002] plane and the long axis of ZnO is estimated to be ca. 10 nm. SWCNTs surround the inner ZnO and the thickness is about 2.5 nm which is equal to the diameter of the two bundle of SWCNTs.

Using HCl, ZnO–SWCNTs are dissolved to remove inner ZnO NPs. And then they were sieved by AAO filter. As shown in Fig. 2.22b, very interesting nano sized and ring shaped SWCNTs, called NR SWCNTs, were obtained. The size of NR SWCNTs is about 20–30 nm and the shape looks like a distorted hexagonal which

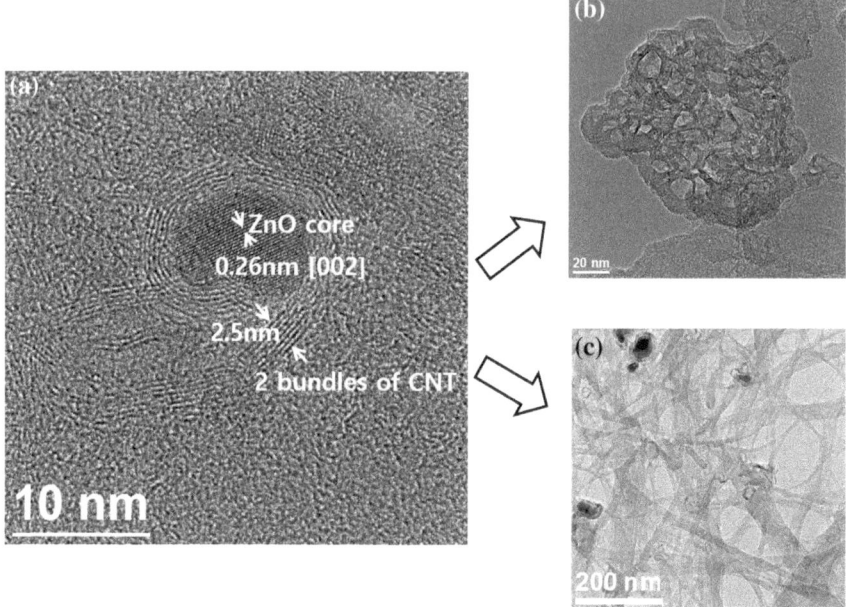

Fig. 2.22 a HR TEM images of ZnO–SWCNTs hybrid structure. After dissolving in HCl (inner ZnO was removed), **b** Obtained NR–SWCNTs with AAO filter and **c** sonificated SWCNTs without AAO filter

may be closely related to the crystalline structure of ZnO NPs. There have been some reports on the formation of ring shape CNTs using several approaches such as CVD, sonification or theoretical calculation (Liu et al. 1997; Martel et al.1999; Sano et al. 2001; Colomer et al. 2003; Komatsu et al. 2006; Guo et al. 2007; Geng et al. 2008; Zhang and Li 2009), the formation of nano ring CNTs with a size of smaller than 30 nm has never reported before. During the formation of the ZnO–SWCNTs hybrid structure, mixed phases of both $Zn(OH)_2$ and ZnO as inner cores are simultaneously observed in a reaction time shorter than 1 h and the boundary between ZnO and SWCNTs seems to be simply rounded like circle. But after the reaction time goes past 1 h, $Zn(OH)_2$ turns into ZnO through dehydration and the number of ZnO NPs on the SWCNTs surfaces increases, which leads to the agglomeration and nucleation. In consequence, ZnO NPs (sizes up to 10 nm) crystallizes into hexagonal structure and this cohesion causes SWCNTs to bend and distort locally. After filtering using anodized aluminum oxide (AAO) ceramic filter (with the hole of 100 nm diameter), very uniform These NR–SWCNTs keep the ring shape even after sonification and after two weeks later. It is believed that this solid NR–SWCNTs, called closed NR–SWCNTs, results from the strong chemical bonding of each carboxyl (COOH) groups induced at the edge of open end in cut SWCNTs as shown in Fig. 2.20. Unfortunately the productivity is very low (<4 %) But after ultra sonification without filtering, most of NR-SWCNTs revert to their

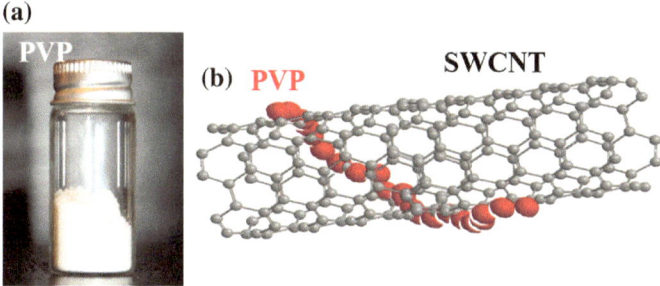

Fig. 2.23   a Polyvinylpyrrolidone (PVP; $C_6H_9NO)_n$) and **b** PVP wrapper bonded with SWCNTs through H-bonding

original linear appearance without maintaining the ring shape as shown in Fig. 2.22c, which implies that a lots of NR-SWCNTs are formed in weak bonding. In order to improve the productivity of NR–SWCNTs, weakly bonded NR–SWCNTs are going to be bound with PVP. It can be easily expected that strong hydrogen bonding between hydroxyl group (–OH) induced SWCNTs and oxygen in $PVP(C_6H_9NO)_n$) and can play a role as a strong wrapper binding between SWCNTs as shown in Fig. 2.23b and keep the ring shape without breaking. By adoption of the PVP wrapper, the productivity of NR-SWCNTs increases up to 98 %.

References

P.O. Anikeeva et al., Nano Lett. **7**, 2196 (2003)
S. Banerjee et al., Adv. Mater. **17**, 17 (2005)
W.J.E. Beek et al., Adv. Mater. **16**, 1009 (2004)
P. Bifeng et al., J. Phys. Chem. C **112**, 939 (2008)
J.F. Colomer et al., Nano Lett. **3**, 685 (2003)
L.J. Cote et al., J. Am. Chem. Soc. **131**, 1043 (2009)
A. Dato et al., Nano Lett. **8**, 2012 (2008)
Z.J. Fan et al., ACS Nano **5**, 191 (2011)
M.G. Frisch et al., *Gaussian 03*, Revision C.02 (2004)
J. Geng et al., J. Phys. Chem. C **112**, 12264 (2008)
A. Guo et al., J. Phys. Chem. C **111**, 3555 (2007)
A.L. Higginbotham et al., ACS Nano **4**, 2059 (2010)
N. Komatsu et al., Carbon **44**, 2089 (2006)
K.N. Kudin et al., Nano Lett. **8**, 36 (2008)
B. Kumar et al., Appl. Phys. Lett. **89**, 071922 (2006)
D. Li et al., Nat. Nanotech. **3**, 101 (2008)
J. Liu et al., Nature **385**, 780 (1997)
J.R. Lomeda et al., J. Am. Chem. Soc. **130**, 16201 (2008)
R. Martel et al., J. Phys. Chem. B **103**, 7551 (1999)
C. Mattevi et al., Adv. Func. Mater. **19**, 2577 (2009)
T.M.G. Mohiuddin et al., Phys. Rev. B **79**, 205433 (2009)
I.K. Moon et al., Nat. Commun. **1**, 1 (2010)

T. Nakajima et al., Carbon **26**, 357 (1988)

Z.H. Ni et al., ACS Nano **2**, 2301 (2008)

A. Palthi et al., J. Macromol. Sci. Part A: Pure Appl. Chem. **47**, 375 (2010)

S. Park et al., Nano Lett. **9**, 1593 (2009)

R.D. Rodriguez et al., Nanoscale Res. Lett. **7**, 682 (2012)

M. Sano et al., Science **293**, 1299 (2001)

D.I. Son et al., Nanotechnology **20**, 195203 (2009)

D.I. Son et al., J. Mater. Chem. **22**, 816 (2012a)

D.I. Son et al., Nat. Nanotechnol. **7**, 465 (2012b)

D.I. Son et al., Nano Res. **5**, 73 (2012c)

L. Tang et al., Adv. Funct. Mater. **19**, 2782 (2009)

G. Wang et al., J. Chem. Phys. **C112**, 8192 (2008a)

F. Wang et al., Science **320**, 206 (2008b)

D. Yang et al., Carbon **47**, 145 (2009)

M. Zhang, J. Li, Materialstoday **12**, 12 (2009)

Chapter 3
Applications of ZnO–Nanocarbon Core–Shell Hybrid Quantum Dots

3.1 ZnO–Graphene and ZnO–C$_{60}$ Ultraviolet (UV) Photovoltaic (PV) Devices

Figure 3.1a, b show the current density-voltage (J-V) curves for the PV devices having ZnO and the ZnO–graphene core–shell QDs as an active UV absorption layer, respectively, fabricated at various speeds from 2000–6000 rpm of spin coating. Using a UV lamp with a wavelength of 365 nm and the incident optical power density of 2 mW/cm^2, the UV photovoltaic performance of Al/Cs$_2$CO$_3$/ ZnO–graphene QD hybrid conjugate/poly- TPD/PEDOT:PSS/ITO device was estimated. The UV PV device with the pure ZnO QD absorption layer fabricated at 4000 rpm exhibited the best performance. The V$_{oc}$, J$_{sc}$, and fill factor (FF) were 0.91 V, 178.3 μA/cm^2, and 0.23, respectively, and the corresponding PCE was approximately calculated as 1.86 %. The other devices prepared at higher or lower speeds than 4000 rpm exhibited relatively lower PCEs of 1.21–1.23 %. On the other hand, the PV device having ZnO–G absorption layer spin-coated at 4000 rpm exhibited a little enhanced PCE of 1.8 %, but the performance of the other devices fabricated at different speeds was significantly enhanced. The best performance was observed for the device fabricated at 5000 rpm and the corresponding values of V$_{oc}$, J$_{sc}$, FF, and PCE were 0.99 V, 196.4 μA/cm^2, 0.24, and 2.33 %, respectively (the maximum power conversion efficiency was estimated based on the methods proposed in the literature (Park and Ruoff 2009). It is revealed that the conjugation of graphene on the ZnO QDs noticeably enhances the PCE from 1.2 to 2.3 %. Large improved performance can connect with the enhanced charge transfer efficiency that results from the highly conducting graphene nanoshells attached to the ZnO core surface; this is evident through a favorable band alignment. In order to understand the underlying phenomena behind the electro-optical behavior of the fabricated UV PV cell, the corresponding energy level diagram (Choulis et al. 2006; Sun et al. 2007; Son et al. 2009; Yang et al. 2010) of the device was constructed and the selected polymer components exhibited a favorable band alignment with the

© The Author(s) 2017
W.K. Choi, *ZnO–Nanocarbon Core–Shell Type Hybrid Quantum Dots*,
Nanoscience and Nanotechnology, DOI 10.1007/978-981-10-0980-8_3

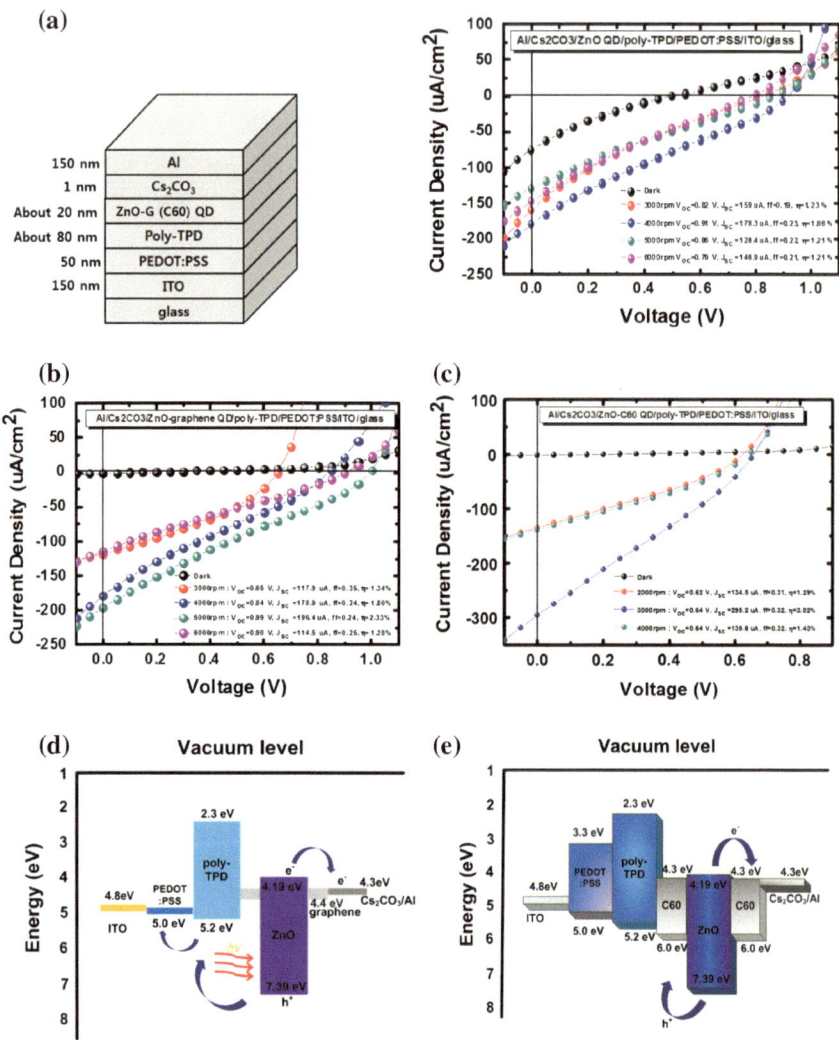

Fig. 3.1 **a** Cross sectional view of ZnO–graphene(ZnO–G) and ZnO–C_{60} based UV photovoltaic device and J-V curve of ZnO only UV absorber PV device. J-V curves of ZnO–graphene (**b**) and ZnO–C_{60} (**c**) devices. Schematic energy level diagram of ZnO–G (**d**) and ZnO–C_{60} (**e**) based UV photovoltaic device (Reproduced from Son et al. 2012a, b)

ZnO–graphene QDs. The energy band diagram for the UV PV device with a forward bias voltage is shown in the left of Fig. 3.1d. Under UV illumination, the ZnO QDs produced photogenerated electron-hole pairs excitons. The chemically conjugated bonding between the ZnO core QDs and the graphene nanoshells facilitated the transfer of the conduction band electrons from the ZnO core QDs to the LUMOs of the graphene nanoshells, resulting in an efficient charge transfer through a hopping mechanism. Since the electron transfer occurs much faster than the radiative and/or nonradiative decay of photoexcitations, the charge recombination process may be significantly hindered. The highly conductive graphene provides direct and efficient pathways for the transport of the conduction band electrons to the Al electrode. However, poly-TPD/PEDOT:PSS may be a hole acceptor, thus assisting in the regeneration of the ZnO QD ground states through a transfer of the hole to the ITO electrode. The efficient transport of the photogenerated holes and electrons led to an enhanced current in the external circuit, which provides high efficiency photovoltaic conversion.

In a similar way, ZnO–C$_{60}$ was also utilized and compared to that device based on ZnO–graphene QDs. As shown in Fig. 3.1c, ZnO–C$_{60}$ Based UV PV device also show the dependence of PCE on the spin-coating speed, i.e. ZnO–C$_{60}$ thickness. As spin-coating speeds are varied from 2000–4000 rpm, Voc's and FF are not altered at the constant value of 0.62–0.64 eV and 0.31–0.32, but the Jsc is the maximum of 295.2 μA at 3000 rpm. Therefore PCE is 1.29 % at 2000 rpm and increases up the maximum of 3.02 % at 3000 rpm, and decreases down to 1.43 % at 4000 rpm. The maximum PCE of 3.02 % for ZnO–C$_{60}$ is greater than that of 2.33 % for ZnO–graphene QDs UV PV devices. The reason why ZnO–C$_{60}$ shows higher PCE efficiency than that of ZnO–graphene is ascribed to difference in the energy level alignment of C$_{60}$ and graphene with poly-TPD (hole transport layer) and Cs$_2$CO$_3$/Al(electron transport layer). Excitons generated from ZnO–(C$_{60}$, graphene) by UV absorption are dissociated into electron and hole pairs by the potential.

Electrons are easily transferred to Cs$_2$CO$_3$/Al electrode due to the energy barrier between the conduction band minimum 4.19 eV (CBM) of ZnO or workfunction 4.4 eV of graphene and LUMO level 4.3 eV of C$_{60}$ and workfunction 4.3 eV of Cs$_2$CO$_3$/Al. In contrast, hole transport is expected to be quite different behavior. Hole injection happened to be quite efficiently from the valence band maximum (VBM) 7.39 eV of ZnO into highest occupied molecular orbital (HOMO) 5.2 eV of poly-TPD through HOMO level 6.0 eV of C$_{60}$ which plays an intermediate energy level. Whereas graphene layer chemisorbed on ZnO inner QDs doesn't contribute to enhance the hole injection by lowering energy barrier between VBM 7.39 eV of ZnO and LUMO level 5.2 eV of poly-TPD due to a quite low workfunction 4.4 eV of graphene.

3.2 An Inverted Polymer Solar Cell Using Functionalized ZnO–Graphene QDs

ZGQDs are well dispersed in relatively polar solvent like ethanol whereas they are quickly precipitated in relatively non-polar solvent like toluene, chlorobenzene, and oxylene. Low dispersion property of ZGQDs in aromatic non-polar organic solvents are correlated with both the presence of relatively large amounts of oxygen-containing groups and also lack of long alkyl chains that can help enhance the solubility in various aromatic solvents. As mentioned in Sect. 1.2, single ZnO

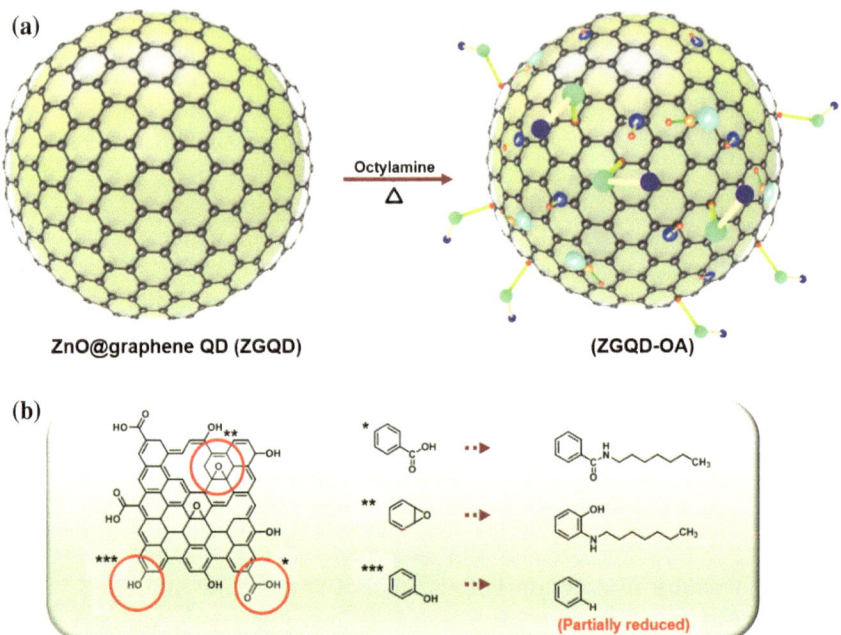

Fig. 3.2 Schematic illustration of chemical modification from ZGQDs to octylamine function-alized ZGQD-OAs (**a**). Surface functionalization of ZGQDs through chemical reactions of graphene nanoshell (**b**) (Reproduced from Moon et al. 2016)

NPs layer or double ZnO/PEI(PEIE) has been widely adopted as high efficient electron transport layer in both QLED and PV devices. However, PEIE layer can be easily damaged by polar organic solvents such as ethanol, IPA and DI-water. Therefore, if PEIE/ZGQDs are used to new kind of ETL double layer, surface of ZGQDs are necessary for functionalization and here is functionalized with octy-lamine (OA) for making well-disperse solution in relatively non-polar organic solvents. The functional alkylamine group will be formed through two chemical reactions at the surface of graphene nanoshell illustrated in Fig. 3.2a. One is the reaction of amine groups with epoxy groups existing on the surface of ZGQDs by way of ring-opening amination of epoxides to yield 1,2-aminoalcohols (Woo et al. 2014; Lee et al. 2014) and the other is the interaction of amines with carboxylic groups along the edge of the ZGQDs to form either amides or alkyl ammonium ions (Mei et al. 2010; Compton et al. 2010). In particular, uniform interface between inorganic ETL layer and active organic layer played an important role in achieving highly efficient photovoltaic cells. Therefore ZnO–graphene core–shell type quantum dots (ZGQDs) was chemically functionalized with octylamine and multi-functionality of the mono-layered ZGQDs as surface modifier, photosensi-tizer and electron transport layer was estimated. ZGQD was prepared by the same method described in Sect. 2.1.1 and then 100 mL of octylamine solution (0.15 M,

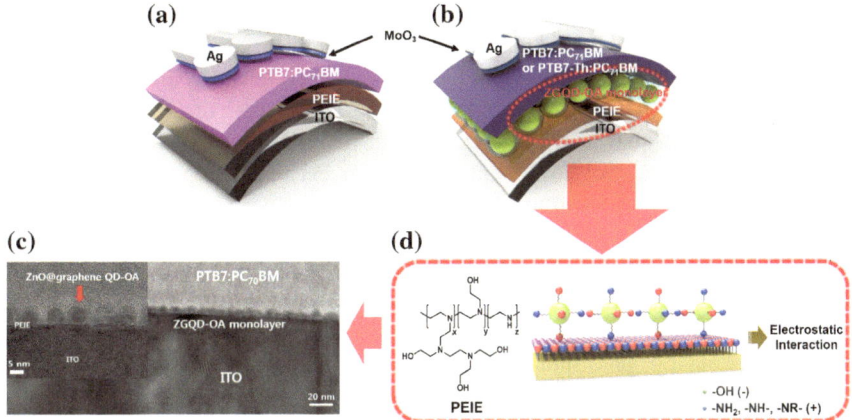

Fig. 3.3 **a**, **b** 3D structure of the inverted PSCs with/without the ZGQD-OAs electron transport monolayer. **c** Cross-sectional TEM image of the inverted PSCs with a ZGQD-OAs monolayer and magnified image of ZGQD-OAs/PEIE/ITOlayer (left inset). **d** Chemical structure of PEIE. The magnified conceptual image of electrostatic chemical interaction between the functional groups of PEIE and ZGQD-OAs (Reproduced form Moon et al. 2016)

in ethanol) was gradually injected into 500 mL of ZGQD solution(5 mg/mL) and stirring at 80 °C for 24 h. After cooling to room temperature, the resulting solution was further centrifuged (Solvents: Hexane, Ethanol and IPA). Finally, the final product was collected after freeze-drying.

An inverted organic photovoltaic solar cell (iPSCs) devices were fabricated on 1.5×1.5 cm^2 pre-patterned ITO coated glass substrates as illustrated in Fig. 3.3. As pre-cleaning and surface treatment steps, the ITO-coated glass substrates were cleaned with ethanol, acetone and 2-propanol. And then the ITO-coated glass was subsequently treated with O$_2$ plasma for 40 s to enhance the surface energy for adhesion and increase workfunction. PEIE solution was spin-coated with <7 nm thickness and the PEIE film then was baked at 110 °C for 10 min under ambient conditions. Afterwards, the OA-ZG interfacial layer was spin-coated as the upper layer of the double structured electron transport layer on the surface of the PEIE layer at 1000 rpm for 40 s, followed by thermal annealing at 110 °C for 25 min. The substrates were then transferred into a nitrogen-filed glove box. The blend of PTB7:PC71BM (or PTB7-Th:PC71BM, 1:1.5 (by weight)) solution were spin-coated onto the PEIE/ZGQD-OAs double layer up to 120 nm thickness as an active absorption layer. As a final step, MoO$_3$ (10 nm) and Ag (80 nm) double structure electrodes as hole transport layer were sequentially deposited on the active layer by thermal evaporation. The as-received solar cells without encapsulation were tested under ambient conditions using a Keithley 2400 SMU and an Oriel xenon lamp (450 W) with an AM 1.5G filter. The photoluminescence lifetimes were carefully measured in time-resolved PL (TRPL) experiments by excitation with 150 fs pulses derived from a 76 MHz amplified Ti:sapphire laser with a wavelength of 750 nm.

From the cross-sectional TEM images of multi-layered inverted PSCs, the surface coverage of a ZGQD-OAs on PEIE/ITO/glass was accurately estimated to be monolayer as shown in the insect of Fig. 3.3c. It is noteworthy that an apparent separation gap was existed between each multilayers of PTB7:PC71BM, ZGQD-OAs, PEIE and ITO. Formation of the ZGQD-OAs monolayer could be well supported by electrostatic interactions either between the positively charged protonated –NH (or –NR) groups of the PEIE and the negatively charged deprotonated –OH groups of the graphene nanoshells, or between the negatively charged deprotonated –OH groups of the PEIE and the positively charged amide (–NH) groups of the graphene nanoshells (Li et al. 2014a, b). In addition, long alkylchains attached to the graphene nanoshells can effectively prevent ZGQD-OAs from vertical stacking by interrupting the intermolecular π-π interactions among graphene nanoshells. This result is also confirmed by AFM topography image taken from ITO, ITO/PEIE, and PEIE/ZGQD-OAs surfaces. The surface morphology does not show much change for ITO and ITO/PEIE whereas it significantly becomes smooth from $\sigma_{rms} = 3.27$ nm (root-mean-square) to $\sigma_{rms} = 2.28$ nm before and after coating the ZGQD-OAs layer on the ITO/PEIE layer. Surface smoothing can effectively decrease both the charge transport distance and contact resistance (Li et al. 2014a, b; Li et al. 2005).

As a consequence, improved physical contact between two layers leads to increment of the short current density J_{sc} (15.6 mA/cm^2→16.2 mA/cm^2) and FF (0.66→0.73), as supported by the decreased series resistance (R_S, 5.0 Ωcm^2→ 3.1 Ωcm^2) and increased shunt resistance (R_{Sh}, 1286.3 Ωcm^2→1554.2 Ωcm^2).

Figure 3.4d shows the J-V curves of the iPSCs fabricated from PTB7:PC71BM blend solutions with/without the ZGQD-OAs monolayer. The iPSCs having ZGQD-OAs layer between PEIE and PTB7:PC71BM showed significant enhancement of 17.8 % PCE from avg. PCE 7.3 to 8.6 % and the champion cell shows a PCE of 8.9 % and the corresponding relative enhancement is 18.7 %. From the results, the PEIE/ZGQD-OAs double electron transport layer is proved to be very efficient for charge generation and transport processes between ITO and active layer (PTB7:PC71BM).

In order to evaluate general suitability of the PEIE/ZGQD-OAs double layer and obtain higher power conversion efficiency, bulk heterojunction (BHJ) iPSCs using PTB7-Th as an electron donor, instead of PTB7 were fabricated. As shown in Fig. 3.4d, control device consisting of ITO/PEIE/PTB7-Th:PC71BM/MoO$_3$/Ag yielded an average PCE of 8.8 %, whereas the PTB7-Th:PC71BM BHJ iPSCs with a combined layer of PEIE and ZGQD-OAs exhibited a higher avg. PCE of 10.0 % (Max.PCE:10.3 %), similar to those of the PTB7:PC71BM system. These performing results give anther strong evidence of PEIE/ZGQD-OAs double layer as a promising electron transport material for high efficiency iPSCs or other photovoltaic devices. For proving the improved performance of iPSCs, the external quantum efficiency (EQE) curves for devices with and without ZGQD-OAs layer were obtained as shown in Fig. 3.4e.

From Fig. 3.4e, the PEIE/PTB7 absorption layer in the wavelength 500–800 nm shows high EQE of 60–80 % due to strong absorption whereas EQE from 300 to

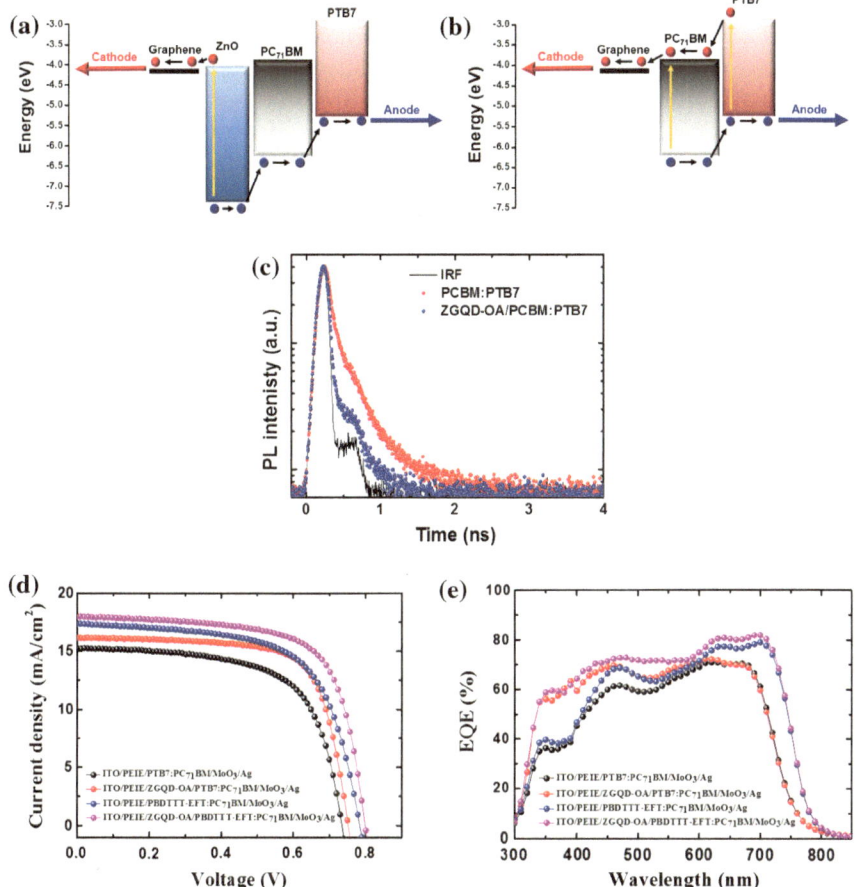

Fig. 3.4 Energy level diagram of the inverted PSCs with (**a**)/without (**b**) ZGQD-OAs monolayer showing pathways for charge generation and transport. TRPL spectra of ITO/PEIE/PTB7: PC71BM (*redline*) and ITO/PEIE/ZGQD-OAs/PTB7:PC71BM (*blackline*) film (λ_{ex}: 420 nm) (**c**). **d** J-V characteristics of the inverted PSCs with/without the ZGQD-OAs monolayer under the illumination of AM1.5, 100 mWcm^{-2}. **e** EQE spectra of the inverted PSCs with/without the ZGQD-OAs monolayer (Reproduced from Moon et al. 2016)

500 nm is relatively weak. But the ZGQD-OAs inserted ITO/PEIE/ZGQD-OAs/PTB7-Th:PC71BM/MoO$_3$/Ag device in the wavelength 350–400 nm exhibits a peak enhancement from 40 to 60 % which originates from due to direct photoexcitation of ZGQD-OAs to the photocurrent in the wavelength range of the UV region (around 380 nm).

Accordingly, the iPSCs of the ZGQD-OAs monolayer between PEIE and PTB7-Th:PC71BM was believed to effectively enhance EQE up to 50 %. This result is supported by the additional spectral responses between 300 and 600 nm. Moreover it is necessary to understand the enhanced charge transport mechanism between the ZGQD-OAs and the PTB7:PC71BM heterojunction with respect to the excited electron–hole recombination. Carrier dynamic behavior can be examined by time-resolved photoluminescence (TRPL) measurements. As shown in Fig. 3.4c, PL life times of the ZGQD-OAs/PTB7:PC71BM structure (blue color curve) decrease significantly in comparison with those of the reference of PTB7:PC71BM structure (red color curve). The calculated average life times at 750 nm for the PTB7:PC71BM and ZGQD-OAs/PTB7:PC71BM structures are 220 and 60 ps, respectively. Moreover, from both TRPL measurements and TEM result, it can be confirmed that the charge transport between PC71BM of PTB7:PC71BM hetero-junction layer and ZGQD-OAs layer in the iPSC devices is through Förster reso-nance energy transfer (FRET) mechanism. From the TEM image in Fig. 3.3c, the distance from the PTB7:PC71BM heterojunction layer to the ZGQD-OAs mono-layer surface is smaller than 10 nm which means that the PTB7:PC71BM hetero-junction layer is located very closely to the surface of ZGQD-OAs monolayer in the iPSC devices enough to occur the charge transfer from PCBM molecules of PTB7:PC71BM heterojunction to the ZGQD-OAs by the FRET process. The FRET process becomes dominant when the donor-to-acceptor distance is smaller than 10 nm (Biju et al. 2006). Thus, the FRET efficiency of the charge transfer can be estimated using the following equation (Biju et al. 2006):

$$\eta = 1 - (\tau_{DA}/\tau_D), \tag{1}$$

where τ_{DA} and τ_D are the fluorescence lifetime of a donor (PTB7:PC71BM) in the presence and in the absence of an acceptor (ZGQD-OAs), respectively. Table 3.1 shows the as-estimated FRET efficiency for the structure with PL lifetimes and the FRET efficiency, which is determined from the time-resolved PL data (73.4 % at 750 nm). The electron decay of the ZGQD-OAs/PTB7:PC71BM structure at selected wavelength (750 nm) was much faster than that for the reference sample of PTB7:PC71BM structure, which indicates the existence of an additional high-efficiency relaxation channel and confirms the efficient electron transfer from

Table 3.1 PL lifetimes of the ITO/PEIE/PCBM:PTB7 and ITO/PEIE/ZGQD-OA/PCBM:PTB7 structures

Film[a]	χ^2	Int.[b]	Amp.[c]	f (%)	τ_1	f (%)	τ_2
With ZGQD-OA	1.06	0.22	0.12	0.44	0.39	0.56	0.08
				0.13	0.39	0.87	0.08
Without ZGQD-OA	1.13	0.06	0.02	0.13	0.31	0.87	0.02
				0.01	0.31	0.99	0.02

Reproduced from (Moon et al. 2016)
[a]Monitored wavelength was 750 nm. The PL decay curves were fitted by a bi-exponential function to calculate the lifetimes of the samples. The intensity weighted average exciton lifetime (τ_{avr}) was $f_1\tau_1 + f_2\tau_2$, where f_1 and f_2 are fractional intensities and τ_1 and τ_2 are lifetimes. χ^2 is the reduced chi-squared value
[b]Intensity (weighted)
[c]Amplitude (weighted)

the PC71BM to the ZGQD-OAs in this case. These observations indicate that inserting ZGQD-OAs electron transport layer can (1) create an additional interfacial region helpful for exciton dissociation to augment the charge transfer rate and (2) prompt the photo-induced charge transfer arising from static quenching and charge-transfer reactions. Additionally, it might (3) act as a hole-blocking barrier caused by introducing deep HOMO energy levels of ZnO while electrons are easily transferred from the PC71BM (LUMO energy level: 4.0 eV) to ZnO (LUMO energy level: 4.1 eV) or graphene nanosheets (WF: 4.3 eV) (Moon et al. 2016).

The workfunction (WF) of both ITO/PEIE and ITO/PEIE/ZGQD-OAs were analyzed using the ultraviolet photoelectron spectroscopy (UPS) as calculating from the secondary electron cutoff as shown Fig. 3.5a. The WF of bare ITO exhibits 4.8 eV and after a thin PEIE layer (ca.5 nm) being coated decreased down to 3.94 eV. The reduction of work function by PEIE surface modifier was known to cause directional formation of dipole and be dependent on the thickness of PEIE (Zhou et al. 2012a, b). The performance of PSCs increased and showed the maximum at 5 nm and then decreased reversely as shown in Fig. 3.5. It is also not possible to increase PEIE thickness due to its insulating nature (bandgap of ~6 eV) (Zhou et al. 2012a, b). It is interesting that adding the second electron transport layer of ZGQD-OAs further decreased WF from 3.94 eV to 3.72 eV without

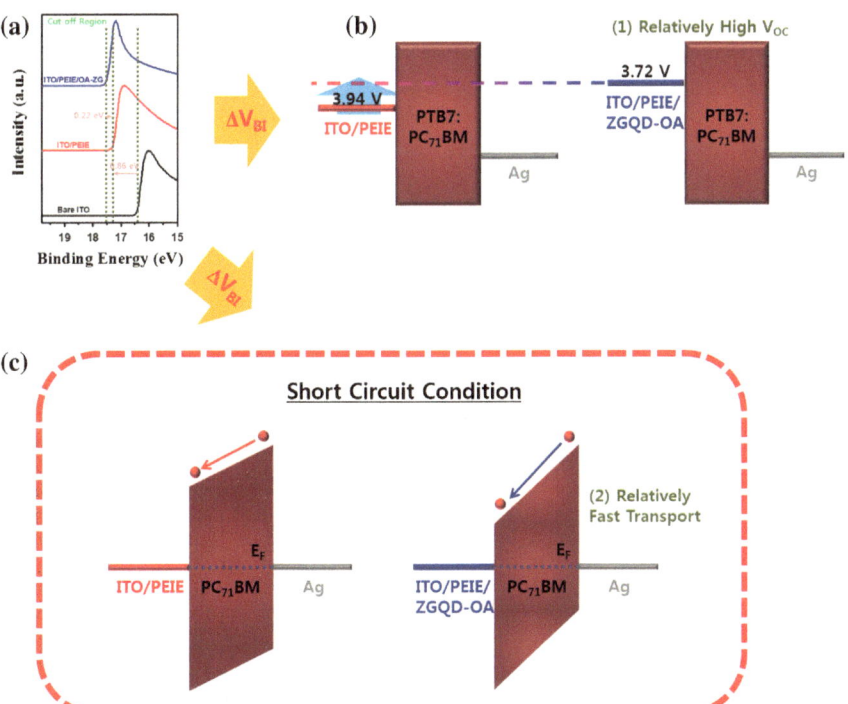

Fig. 3.5 Energy level diagram of the inverted PSCs with (**a**)/without (**b**) ZGQD-OA (Reproduced from Moon et al. 2016)

Fig. 3.6 Device performance
of polymer solar cells with
various PEIE layer
thicknesses (Reproduced from
Moon et al. 2016)

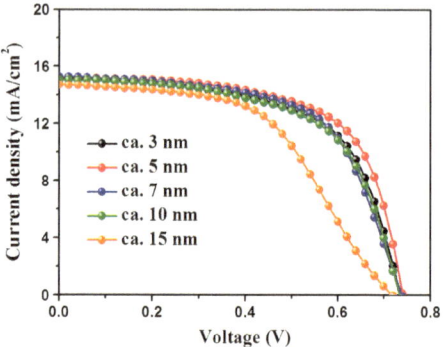

increasing PEIE thickness (Fig. 3.5b).The increased built-in voltage (V_{BI}) due to
the increased WF contributes to (1) the improved V_{oc} (from 0.73 to 0.75 V,
Fig. 3.5c), also, (2) larger band bending at the PC71BM surface under short circuit
conditions (Fig. 3.5c). Thus the electrons can be effectively transported to the ITO
electrode, resulting in the enhanced J_{sc} (from 15.6 to 16.2 mA/cm^2) and FF (from
0.66 to 0.73) (Fig. 3.6).

3.3   ZnO–Graphene Core–Shell Hybrid QDs Light Emitting Diodes

3.3.1   White Light-Emitting Diode

The multilayer structured LED device was fabricated by using wet chemical
spin-coating method using a ZnO–graphene quantum-dot solution as the emissive
layer (Son et al. 2012a, b). As an anode substrate, a commercially available
laser-patterned transparent conducting indium tin oxide (ITO) glass with a sheet
resistance of Rs = 15 Ω/sq^2 was used. Before spin-coating, the ITO glass was
cleaned by a routine chemical procedure including sonication in a detergent (acetone, methanol and isopropyl alcohol solution), rinsing with deionized water and
then treatment with ultraviolet ozone in a chamber for 15 min. The PEDOT:PSS
polymer in C_3H_7OH (*i*-ProH) (2.39 wt%) solution was sonicated for 15 min and
spun onto the ITO/glass as an hole injection layer (HIL) and then thermally treated
at 100 °C on a hot plate for 30 min. The poly(N,N′-bis(4-butylphenyl)-N,N′-bis
(phenyl)benzidine) (poly-TPD) polymer in chlorobenzene (1.5 wt%) solution was
sonicated for 15 min and spun on PEDOT:PSS/ITO/glass as a hole-transport layer
(HTL), then annealed at 120 °C for 20 min on a hotplate. The ZnO–graphene
quantum-dot powder in C_2H_5OH solution (3.07 wt%) was sonicated for 15 min and
spun onto the poly-TPD/PEDOT:PSS/ITO/glass as the active layer of the LED,
then annealed for 20 min on a hotplate at 110 °C. Cs_2CO_3 in 2-ethoxyethanol
($C_4H_{10}O_2$) solution (0.01 wt%), after sonication for 15 min, was spun onto the

ZnO–graphene quantum-dot layer as the electron injection layer (EIL) and hole blocking layer (HBL). Subsequently, the film was annealed at 90 °C for 10 min to remove the solvent. Afterwards, as a top cathode, an Al layer with a thickness of 150 nm was deposited through a shadow mask by thermal evaporation. The effective emission area of the LEDs was 2 mm × 2 mm.

Figure 3.7a shows the J-V characteristic curve of a ZnO–graphene quasi-core–shell quantum-dot LED as the applied voltage varies up to 17 V. With increasing applied voltage, J remains approximately constant (increasing slightly) up to ∼11–12 V, then increases drastically which corresponds to the turn-on voltage of the LED. Figure 3.7b shows the electroluminescence (EL) at different applied voltages of 11–17 V. Four distinctive EL peaks centered at 428 nm (2.89 eV), 452 nm (2.74 eV), 475 nm (2.60 eV) and 606 nm (2.04 eV) were observed respectively. All the peak positions slightly differ from those of the photoluminescence. In the energy band diagrams for the ZnO–graphene quantum-dot LED (Fig. 3.7c), Cs_2CO_3/Al is used as both an electron-injection layer (EIL) and a hole-blocking layer (HBL) (deposited by a chemical method) rather than the widely used thermally evaporated LiF (Huang et al. 2007) because it

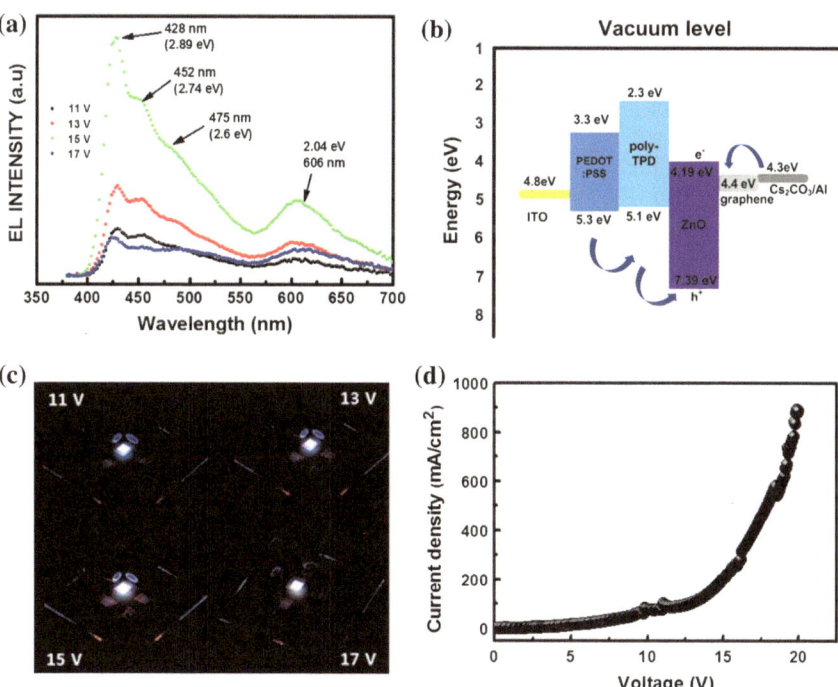

Fig. 3.7 **a** Electroluminescence spectra of the fabricated ZnO–graphene quasi-quantum dot LED device with applied voltages from 11 to 17 V. **b** Band diagram of the fabricated LED device. The pathways of holes and electrons are indicated by arrows. **c** Photograph of light emission at 11, 13, 15 and 17 V applied bias, respectively. **d** Current density-voltage (J-V) characteristics for the fabricated LED device (Reproduced from Son et al. 2012a, b)

(4.3 eV) well aligns with the energy level of graphene (4.4 eV). The HOMO and LUMO levels of PEDOT:PSS, poly-TPD, graphene and ZnO are taken from the literatures (Huang et al. 2007; Anikeeva et al. 2007; Sun 2008; Son et al. 2009). Electrons are injected from the Cs_2CO_3 to the LUMO and LUMO+2 excited MOs of the G-O_{epoxy} conjugated with the ZnO quantum dot rather than to the conduction band of the ZnO (Fig. 3.7c), because the 4.4 eV energy level of graphene (Yang et al. 2010) is lower than the 4.19 eV conduction band of ZnO. In the creation of the LUMO and LUMO+2 MOs of G–O_{epoxy}, the first two emissions, 428 nm (2.89 eV) and 452 nm (2.74 eV), can be easily ascribed to excitonic recombination of injected electrons on the unoccupied LUMO and LUMO+2 MOs of G–O_{epoxy} and holes in the valence band of ZnO. Under forward bias, electrons injected from the Cs_2CO_3/Al electrode result in a net carrier doping, that is, an upward shift of the Fermi energy (E_F) of the graphene (Fig. 2.15). Accordingly the valence band of ZnO has to also be slightly lifted to align the Fermi level of ZnO with that of G–O_{epoxy}. This shift makes the energy differences between the unoccupied LUMO and LUMO+2 MOs of G–O_{epoxy} and the valence band of ZnO reduced, so the peak positions of the two emissions shows redshift. The applied voltage V changes the charge-carrier density n in graphene ($n = \alpha V$), and therefore shifts the Fermi level E_F, where $\Delta E_{F=}$ $\hbar v_{F(}\pi|n|)^{1/2}$. Here, positive (negative) n means electron (hole) doping, v_F is the Fermi velocity (0.8×10^6 m s^{-1}) and $\alpha \approx 7 \times 10^{10}$ cm^{-2}V^{-1} (estimated from a simple capacitor model) (Wang et al. 2008). The shift in E_F to align with that of graphene affects the positions of the VB and CB of ZnO. For an applied voltage of V $\approx$ 11–15 V, the calculated ΔE_F is as much as $\sim$82–95 meV and the Fermi level of graphene is raised to the same level. Assumed ΔE_F to be 95 meV for V = 15 V, EL are expected at 2.95 and 2.80 eV, respectively which agree well with the observed electroluminescence values, 2.89 eV (428 nm) and 2.74 eV (452 nm) qualitatively. Therefore it is reasonable that the change in the peak position of EL can be attributed to the shift of E_F due to the injection of electrons. The other EL emissions at 475 nm (2.60 eV) and 606 nm (2.04 eV) can be explained as follows: when forward bias is applied, blue-light emissions occur by recombination of holes in the valence band of ZnO with electrons in the s orbital of the G–O_{epoxy}-related LUMO and LUMO+2 MOs (Fig. 3.7c). These blue emissions are then reabsorbed in poly-TPD and the corresponding newly generated electron-hole pairs recombine and produce another blue emission at 475 nm. Subsequently, for a PEDOT:PSS layer, the electrons in the HOMO level of PEDOT:PSS are also excited into LUMO level by emission from the poly-TPD layer. The observed emission from this layer, at 606 nm, coincides exactly with the bandgap (2.0 eV) of PEDOT:PSS.

Figure 3.7d presents a photograph of the light emission from our fabricated LED for applied biases of 11, 13, 15 and 17 V, respectively. The ZnO–graphene quantum-dot layer LED pixels appear uniformly luminescent and look bluish-white to the naked eye due to the combination of a series of blue emissions (428, 452 and 475 nm) and yellow emission (606 nm), with CIE coordinates (0.23, 0.20), (0.28, 0.24) and (0.31, 0.26) for 13, 15 and 17 V applied biases, respectively. At 15 V applied bias and with optimal CIE coordinates (0.23, 0.20), the maximum luminance reaches ca.798.1 cd/m^2. The external quantum efficiencies of the fabricated

LEDs were measured to be 0.18, 0.04 and 0.02 % for 13, 15 and 17 V applied biases, respectively (Son et al. 2012a, b).

3.3.2 Optimization

In order to further improve the performance of ZnO–graphene QDs based LED, mainly surface treatment and the thickness of HTL layer among several processes were slightly controlled simultaneously along with different annealing temperature and plasma gas species as listed in Table 3.2.

Firstly ITO surface was pretreated by O_2 plasma instead of N_2 plasma for increasing both the workfunction of ITO from 4.8 eV up to 5 eV and this made ITO surface very smooth and hydrophilic useful for successive spin-coating process. PEDOT:PSS hole injection layer was treated with the same method, but poly-TPD layer was spin-coated at 5000 rpm and annealed at further elevated temperature 130 °C which means thickness of poly-TPD became thin. In case of spin-coating of ZnO–graphene QDs, only annealing was changed from 1 h to shorter time 30 min. EIL of Cs_2CO_3 layer was also annealed at slightly elevated temperature at 130 °C. It is noteworthy that the applied voltage after optimization was much reduced. As shown in Fig. 3.8a, electroluminescence was largely increased at the applied voltage 10 V and blue emission centered at 423, 451, 477 nm and yellow emission at 611 nm were much enhanced together. From Fig. 3.8b, after optimizing the fabrication conditions of ZnO–graphene hybrid QDs based LED devices, luminance was remarkably enhanced from 789 cd/m^2 at 15 V to 1118 cd/m^2 at 10 V, which was 140 % increase at even lower voltage. Before optimization, only external quantum efficiency except efficacy was measured and only as much as 0.04 % at 15 V. Efficacy was much improved as 0.15 cd/A respectively and can be expected to be further increased by controlling the thickness of EIL and ETL. CIE (Commission International de l'éclairage) coordinate (x, y) = (0.23, 0.20) for conventional device was also changed into (x, y) = (0.31, 0.26) after optimization. This result can be thought as resulting from the increase of both hole injection between ITO and PEDOT:PSS and hole-electron recombination rate by the control of thickness of HTL.

Table 3.2 Conventional and optimization condition for the improvement of ZnO–graphene QDs-based LED

	Conventional	Optimization
Plasma treatment	N_2	O_2
PEDOT:PSS	4000 rpm, 110 °C, 30 min	4000 rpm, 110 °C, 30 min
Poly-TPD	4000 rpm, 110 °C, 1 h	5000 rpm, 130 °C, 1 h
ZnO–graphene	4000 rpm, 110 °C, 1 h	4000 rpm, 110 °C, 30 min
Cs_2Co_3	5000 rpm, 110 °C, 1 h	5000 rpm, 130 °C, 1 h
EQE (%) or efficacy (cd/A)	0.04 %	0.15 cd/A
Luminance	798 cd/m^2@15 V	1118 cd/m^2@10 V
CIE	(x, y) = (0.23, 0.20)	(x, y) = (0.31, 0.26)

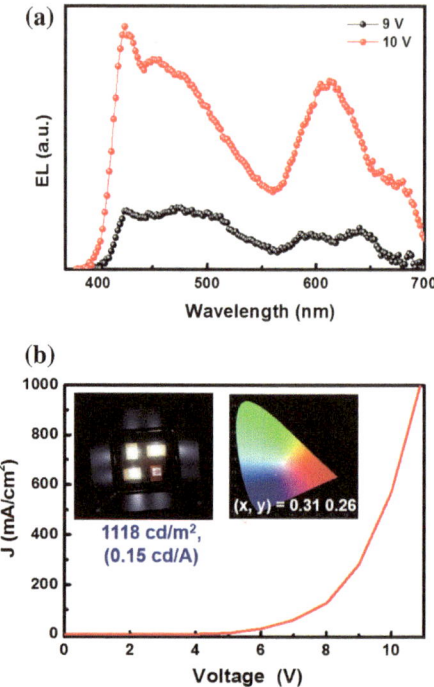

Fig. 3.8 **a** Electroluminescence and **b** J-V curve for ZnO–graphene hybrid QDs based LED fabricated at the optimized conditions (Insets are photographic images of emission from the device and CIE coordinate)

3.4 Passive Matrix ZnO–graphene Core–Shell Hybrid QDs Light Emitting Diodes

3.4.1 Glass Substrate

As shown in Fig. 3.9a, passive matrix ZnO–graphene-based QD LED with 10 × 10 pixels was fabricated on a 5 cm × 5 cm ITO glass. And then this device was programmed by external circuit and was tested to see whether each pixel in passive mode really worked. Figure 3.10 illustrates the performance of passive matrix ZnO–graphene hybrid QDs-based LED. From the left, single dot (1 × 1), line (1 × 10), and matrix mode (10 × 10) mode are all successfully achieved at the operational voltages of 8.0–9.8 V. Also by using the programmed ZnO–graphene hybrid QDs-based LED, initial logo of K-I-S-T was displayed in succession as depicted in Fig. 3.11.

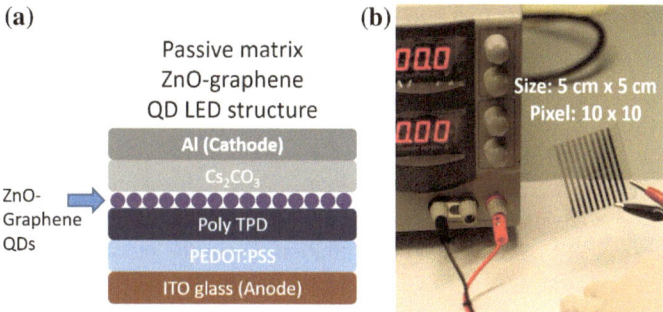

Fig. 3.9 Passive matrix ZnO–graphene hybrid QDs-based LED (**a**) Cross sectional view of the QLED and (**b**) photographic real image of the 5 cm × 5 cm QLED with 10 × 10 pixels

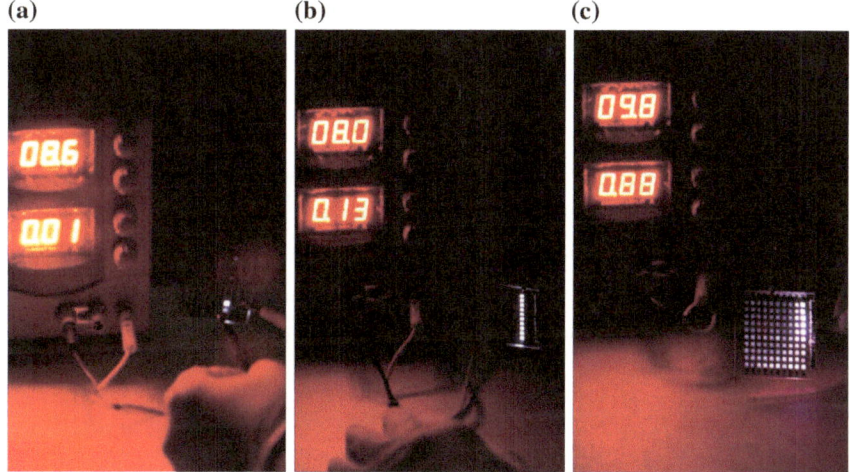

Fig. 3.10 Passive matrix ZnO–graphene hybrid QDs-based LED, **a** single dot mode, **b** One line (1 × 10 matrix) mode, and **c** 10 × 10 matrix mode

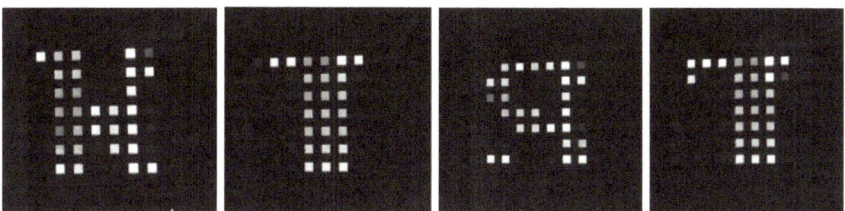

Fig. 3.11 Logo initial of K-I-S-T is displayed by programmed passive matrix ZnO–graphene hybrid QDs-based LED

3.5 Flexible Substrate

In a similar way, passive matrix ZnO–graphene-based QD LED with 10×10 pixels was also fabricated for the first time on flexible 5 cm × 5 cm ITO/PET substrate as shown in Fig. 3.12a. At the operational mode, very bright white emission was observed in matrix mode from the device (1 line did not fully worked). And even at the bending stage, white emission continued without any change in intensity and substrate configuration. From the result, it is believed that ZnO–graphene-based QD LED is quite suitable for flexible planar lighting. Considering the state of the art in OLED, this is very promising because despite the production of flexible OLED in large scale will not be easily established in the near future, QD-based large area production is rather convenient and quite feasible on the flexible substrate at a moderate temperature.

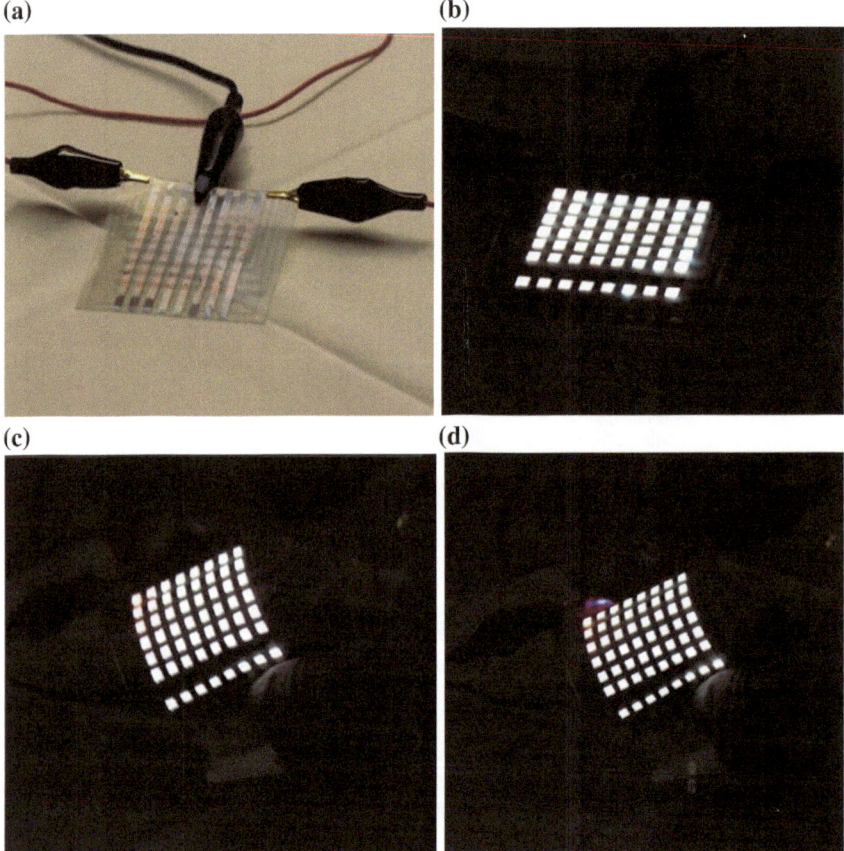

Fig. 3.12 Flexible passive matrix ZnO–graphene hybrid QDs-based LED fabricated on ITO/PET substrate. **a** Real images of the device, white light emission from the device without bending (**b**) and with bending (**c**, **d**) degradation image occurred at the edge at high voltage mode

Fig. 3.13 Deposition of active emitting layer of ZnO–graphene hybrid QDs by using slot-die coater (*Arrow* indicates 5 cm × 5 cm PET/ITO/PEDOTT:PSS/poly-TPD substrate paced on 25 cm × 25 cm glass pad)

For the large scale fabrication in the solution process, deposition of the active layer of ZnO–graphene hybrid QDs was carried out by using the batch type slot-die coater with effective deposition area (250 mm in width) as shown in Fig. 3.13. Prior to coating ZnO–graphene hybrid QDs layers, hole injection layer PEDOT:PSS and hole transport layer poly-TPD hole were already spin-coated on the patterned 5 cm × 5 cm ITO/PET substrate. And then it was patched on the 25 cm × 25 cm glass pad substrate (indicated by white arrow) and placed on carrier stage in slot-die coater. At the optimum slot-die gap and coating speed, the ZnO–graphene hybrid QDs layer ink was coated by slot-die and then dried at 90 °C for 1 h.

3.6 ZnO–Graphene and ZnO–C$_{60}$ Photoelectrochemical Devices

Since the metal oxide materials have good environmental and thermal stability, especially metal semiconductors could have been widely applied for electrode in harsh environmental energy devices, for instance, anode for lithium-ion secondary battery (LIB) (Kim et al. 2005; Park et al. 2009), carrier transport layer for dye-sensitized solar cells (DSSCs), and photoanodes in a photoelectrochemical (PEC) which is a solar-to-chemical energy conversion. In case of PCE photoandoe, they can take part in the reaction with components in the electrolyte during water photo-oxidation, which deteriorate their long term stability and lowers photocurrent-to-oxygen conversion faradaic efficiency (Krol et al. 2008; Paracchino et al. 2011). In particular, among the metal oxides, Hematite α-Fe$_2$O$_3$ was firstly adopted in 2000 (Beerman et al. 2000) as a PEC photoanode due to chemical stability and favorable band-gap 2.1 eV besides abundance in nature and inexpensive cost. But a very short life time in excited state (~ 10 ps) (and a short hole diffusion length (~ 2–4 nm) seriously limit the efficiency of its charge separation and collection as a PCE anode. On the other hand, TiO_2 and ZnO have been emerged new candidate for photoanode because TiO_2 has long hole diffusion length (~ 10 nm) (Salvador 1983) and electron mobility (~ 1 cm^2/Vs) (Hendry et al. 2004) and ZnO has 10–100 time higher electron mobility than that of TiO_2. However, although ZnO has outstanding carrier diffusion length, photoelectric conversion and higher photocatalytic dye degradation efficiency (Sohn et al. 2006) than TiO2, it has been difficult for ZnO to surpass TiO_2 because of its poor stability in liquid electrolytes and acid environments (Wang et al. 2011; Chemelewski et al. 2014). Unfortunately in the case of a ZnO photoanode, oxygen atoms in the ZnO lattice rather than H_2O molecules are easily oxidized by photogenerated holes that produce Zn^{2+} under band gap excitation as shown in Fig. 3.14b, a process called

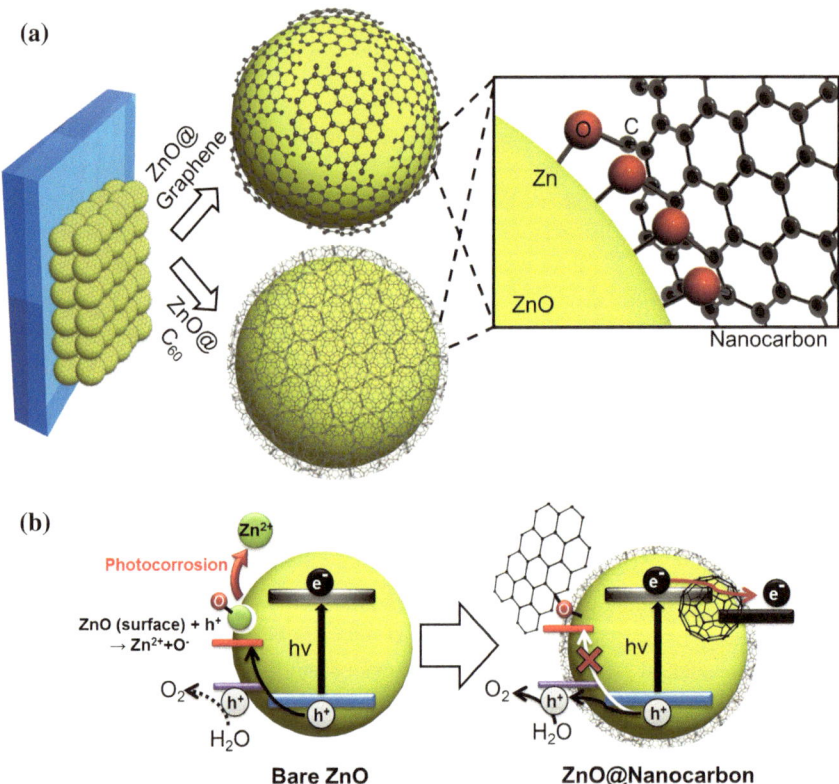

Fig. 3.14 a Schematic of photoanode composed of ZnO–graphene and ZnO–C_{60} nanoshells, and the enlarged image of chemical bonding between the functional groups and Zn^{2+} (Zn–O–C bonding). **b** Schematic diagram of water oxidation mechanism of ZnO-nanocarbon core–shell QDs (Reproduced from Kim et al. 2015)

the photo-corrosion phenomenon, because the oxidation potential of ZnO is more negative than that of H$_2$O molecules, analogous to cadmium complexes such as CdS or CdSe (Hill and Choi 2012). To oppress the drawbacks caused by the surface state, the components in the electrolyte were optimized through surface modification by various catalyst, deposition of overlayer, and addition of heterojunction structures (Kudo and Miseki 2009; Zhong et al. 2011; Yanh et al. 2013; Kim et al. 2013). It was previously reported that the photogenerated electrons in a semiconductor could be comfortably transferred into graphene sheets by forming semiconductor/graphene nanocomposites or semiconductor particles wrapped with graphene sheets like ZnO with GO (graphene oxide) and CdS with graphene (Bolotin 2008; Cao et al. 2010; Li et al. 2012). Because of its good charge conductance, the charge separation efficiency of the photogenerated electrons and holes could be enhanced so as to improve the PEC response. Nevertheless, despite many efforts to manipulate more uniform semiconductor-graphene nanocomposite structures by wrapping reduced graphene oxides on semiconductors (Luan et al. 2013), or inserting semiconductors between graphene sheets (Zhou et al. 2012a, b), the semiconductors were still randomly scattered and nonuniformly attached to the graphene-based sheets. In these non-ideal nanostructures, the dissociated electrons and holes from semiconductors could be quenched or recombined into the graphene sheets before being collected in the electrode or participating in the water oxidation reaction.

As described in Sects. 2.1.1 and 2.2.1, quasi-core–shell quantum dot (QDs) structures composed of ZnO–graphene and ZnO-fullerene (C$_{60}$) were prepared as shown in Fig. 3.14 using a simple chemical method. Charge transfer from the photogenerated electrons excited in the valence band from conduction band in ZnO core are occurred efficiently into the nanocarbon (graphene QDs and fullerene) shells with high electron conductivity and the remaining holes in the valence band can participate in the oxidation reaction. The strong oxygen bridges between the ZnO core and the nanocarbons not only permit ultrafast photogenerated electrons for enhanced photocatalytic activity, but also greatly enhance the long-term stability of ZnO.

The photoelectron chemical response of the core–shell structure was investigated as shown in Fig. 3.15, using a three electrodes system in 0.5 M NaClO$_4$ electrolyte (pH 6.9). Figure 3.15a represents the photocurrent density-potential (J-V) curves ranging from ~ 0.4 to 0.6 V versus Ag/AgCl reference electrode (equivalent to 0.23–1.23 V vs. reversible hydrogen electrode (RHE)) under the chopped 1 sun illumination (AM1.5G, 100 mW/cm^2). Figure 3.15b shows the J-V curves behavior under 1 sun illumination without chopping, which is coincident with Fig. 3.15a (with chopping). The bare ZnO QD photoanode as a reference sample represented very poor photocurrent density and this is because the photo generated holes in the ZnO are easily consumed by the self-oxidization of ZnO with oxygen atoms on the surface of ZnO (Hill and Choi 2012). Compared to ZnO only reference, the core–shell ZnO–graphene and ZnO–C$_{60}$ photoanodes showed greatly enhanced

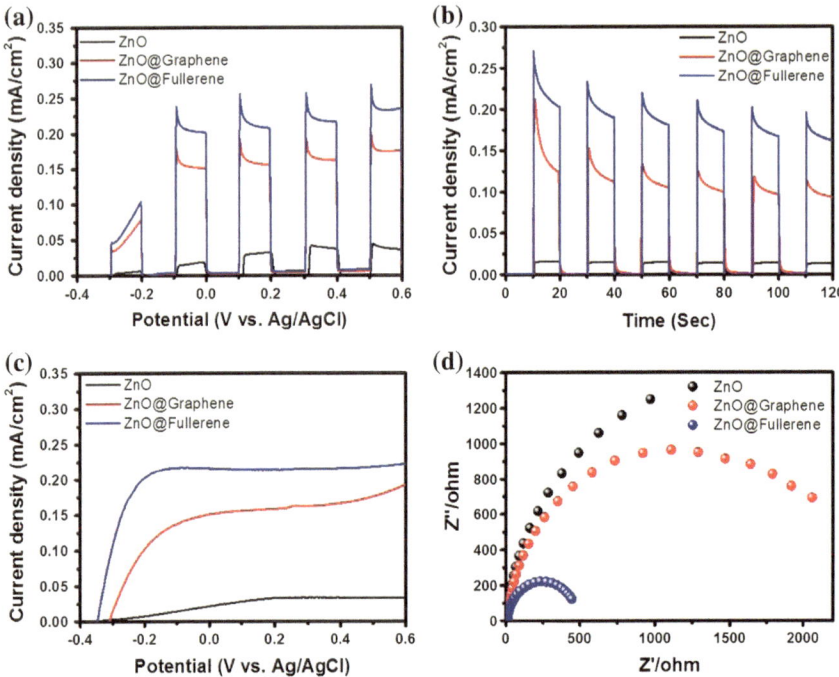

Fig. 3.15 a Current density to potential (J-V) curves and **b** current density to time curves under on/off chopped light using 1 sun light source (100 mW/cm^2), **c** J-V curves obtained by linear sweep voltammetry (LSV) under 1 sun irradiation, and **d** Nyquist plot of electrochemical impedance spectroscopy (EIS) obtained by using 0.5 M NaClO$_4$ electrolyte (pH 6.9), a saturated Ag/AgCl reference electrode, and a Pt counter electrode (Reproduced from Kim et al. 2015)

photocurrent density under exactly the same experimental conditions. At 0.6 V versus Ag/AgCl (1.23 V vs. RHE), photocurrent densities of 0.24 mA/cm^2 for ZnO–C$_{60}$, 0.18 mA/cm^2 for ZnO–graphene, and 0.04 mA/cm^2 for bare ZnO were obtained respectively. Furthermore, Fig. 3.15b shows that the photocurrent density of ZnO–C$_{60}$ photoanode as a function of response time much sharply increase at its on-set potential comparing to that of both ZnO–graphene and bare ZnO photoanodes, where the slope of J-V curve is closely dependent on the resistance and the charge transport properties of the photoanode (Bard and Faulkner 2001). From Fig. 3.15d, the charge transport resistance resulting from preventing the photo generated carriers from being consumed by a charge recombination or lose is decreased and which reversely increases charge collection at the electrode and improvement of the current density near to the on-set potential. The photocurrent enhancement by replacing of ZnO–graphene and ZnO–C$_{60}$ photoanodes implies that photo generated electrons were conveniently transported from the ZnO core to the graphene or C$_{60}$ shells and photo-corrosion is much weakened; consequently these led the photo induced charge carrier separation efficiency to be much improved. Because the energy level of reduction states of the graphene and the fullerene are moderately placed between the conduction band (CB) of ZnO and the WF of the transparent conducting oxide (TCO) substrate, electrons are efficiently extracted from ZnO.

In view of electrochemistry, first, the CB of ZnO is −0.5 V versus normal hydrogen electrode (NHE) and the reduction state of graphene were known as −0.3 V versus NHE at pH7 respectively (Sher Shah et al. 2013; Li et al. 2014a, b). Delocalized electrons in the graphene and oxygen atoms in ZnO bridging the ZnO QDs and graphene can provide a sufficient transport passage of photo-excited electrons (Xiao et al. 2014). Second, as for the ZnO–C$_{60}$, not only the delocalized conjugated electrons but also −0.2 V versus NHE of the lower reduction state of fullerene facilitates the photo induced electron transport from ZnO to fullerene and the TCO substrate (Kamat 1991, 1994). The sufficient conjugated electrons and good electron taking away property of C$_{60}$ could improve the electron injection from ZnO to fullerene (Rao and Voggu 2010). Moreover, due to the much efficient reduction state of fullerene in comparison to graphene, ZnO–C$_{60}$ photo anode shows significantly better enhanced PEC response than ZnO–graphene anode. Besides, the charge trap or recombination sites were complemented by the nanocarbon shells with the oxygen bridge bonds, which resulted in efficient charge flow from ZnO to TCO substrate. In Fig. 3.15c, the linear sweep voltammetry (LSV) result reveals that the onset potentials of ZnO–graphene and ZnO–C$_{60}$ in photocurrent density curves are negatively shifted (ZnO–C$_{60}$ is more shifted) compared to that of bare ZnO. Due to much more proficient cascade electron transport pathway from the ZnO core to the C$_{60}$ shell comparing to the graphene shell induced by the aforementioned reduction potential states, the onset potential of ZnO–C$_{60}$ is further negatively shifted than ZnO–graphene. The strong coupling between ZnO–graphene or C$_{60}$ greatly reduced the charge transfer resistance and also the flat band potential values of ZnO–graphene and ZnO–C$_{60}$ was well defined by the reduction states of graphene and C$_{60}$ have more negative potential than the CB of ZnO. The modified values were estimated by Mott-Schottky plots: 0.402 V for bare ZnO, 0.339 V for ZnO–graphene and 0.295 V for ZnO–C$_{60}$ versus RHE in 0.5 M NaClO4 electrolyte (pH 6.9) (Kamat 1991).

Furthermore, as shown in a Nyquist plot of electrochemical impedance spectroscopy (EIS) (Fig. 3.15d), the semicircles of ZnO–graphene and ZnO–C$_{60}$ photo anodes were much smaller than that of the bare ZnO photoanode, which signified that the core–shell structures had significantly lower resistance than the bare ZnO structure. The radius of ZnO–C$_{60}$ was much smaller than that of ZnO–graphene which indicates that charge transport resistance (R$_{ct}$) for ZnO–C$_{60}$ was much reduced due to the efficient charge transport. By fitting the EIS results using Z-Viewer software, the R$_{ct}$ values for ZnO–graphene and ZnO–C$_{60}$ were calculated

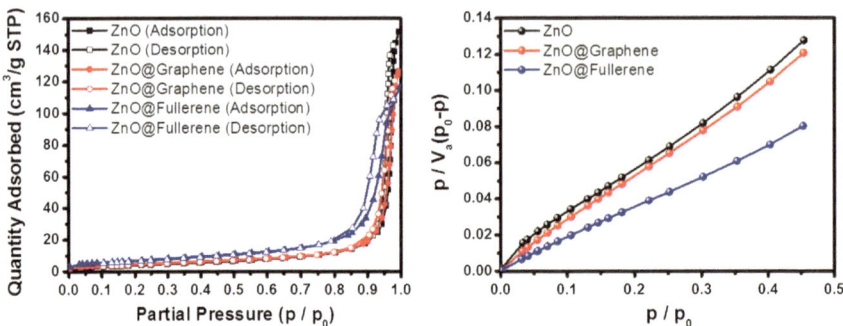

Fig. 3.16 BET surface area analysis of bare ZnO, ZnO@Graphene and ZnO@C_{60} core–shell QDs (Reproduced from Kim et al. 2015)

as 1945 and 474 Ω respectively. These EIS results infer that the photo induced electrons and holes were rapidly dissociated and separated for water oxidation, thereby greatly improving the PEC response (Ma et al. 2014).

As shown in Fig. 3.16, the BET specific surface area of ZnO–graphene (16.906 m^2/g) was almost similar to that of bare ZnO (16.226 m^2/g), which implies that the inner ZnO QDs surface was quite conformally covered by graphene shells, whereas the specific surface area of ZnO–C_{60} (25.287 m^2/g) was much larger due to additional voids formed by the fullerene C_{60}. Therefore water feels larger surface area and can contact the ZnO surface to be split through these extra voids.

To investigate photocatalytic activity of these devices, a photo-degradation experiment using Rhodamine B (Rh.B) was systematically carried out. 20 mg of the powder samples were mixed in 40 ml of 10^{-5} M Rh.B aqua solution. After 2 h stirring in dark conditions to achieve an equilibrium state between adsorption and desorption, AM 1.5 G of light was irradiated. A fixed quantity of Rh.B solution was extracted during the same time interval for every sample (Kim et al. 2011; Mao et al. 2012; Son et al. 2012a, b). Figure 3.17a shows photocatalytic activity of the core–shell QD samples. By using the $C/C_0 = \exp(-kt)$ equation where C_0 is the initial concentration when C is the concentration after 10 min irradiation, rate constants (k)(min^{-1}) are calculated. Rate constant for ZnO was merely 0.0088 and slightly increased to 0.053 for ZnO–graphene. ZnO–C_{60} shows a k value more than 10-times higher than bare ZnO because the charge transportation property of ZnO–C_{60} is much higher. Figure 3.17b shows the results of long-term stability tests performed by repeating the Rh.B photo decomposition experiment using bare ZnO and ZnO–graphene QDs. As shown in Fig. 3.18, normalized photocurrent density to time curves demonstrate that the ZnO–graphene core–shell structures have better improved long-term stability performance than ZnO–C_{60} because 2-dimensional graphene can conformally cover the surface of the ZnO QD core better than

Fig. 3.17 a Concentration (C) changes of Rhodamine B (Rh.B) aqueous solution from initial concentration (C$_0$) as a function of the 1 sun irradiation time and **b** photodegradation stability test. The sample powders were collected and re-dispersed in new Rh.B solution with a time interval; 80 min interval for ZnO and 40 min interval for ZnO–graphene (Reproduced from Kim et al. 2015)

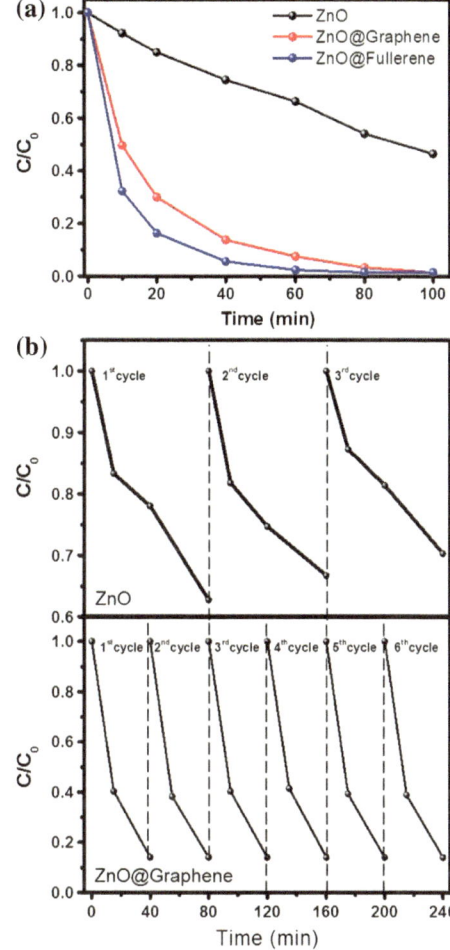

Fig. 3.18 Normalized current density to time curves at 0.5 M NaClO$_4$ electrolyte (pH 6.9) with 0 V (vs. Ag/AgCl reference electrode) of bias under chopped 1 sun irradiation (Reproduced from Kim et al. 2015)

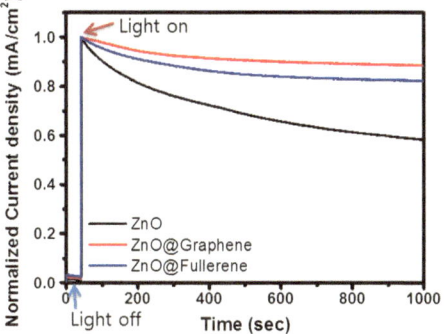

0-dimensional C_{60} can. The enhanced long-term stability of the ZnO–graphene and ZnO–C_{60} core–shell structure was evaluated using chronoamperometry. Each degradation test of pure ZnO was performed for 80 min, but the test of ZnO–graphene or ZnO–C_{60} powder was performed for 40 min because of its higher photocatalytic activity. The strong oxygen atom bridges connecting the ZnO core and graphene shell tremendously improved the long-term stability of the ZnO QDs.

3.7 PVDF-TrFE Composite with NR SWCNTs

Piezoelectric polymers have advantages over piezoelectric ceramics for certain applications wherein acoustic impedance similar to that of water or living tissue is required (Gregorio and Ueno 1999). Also they will be suitable for flexible energy harvesting piezoelectric devices. Due to this reason PVDF is increasingly used for medical and industrial applications.

Copolymer P(VDF-TrFE) resins (Piezotech Co.) in powder or pellet form (composition: 70 %/30 % in mole (VDF/TrFE)) are used. P(VDF-TrFE) copolymers exhibit tailorable ferroelectric, piezoelectric and structural properties that may be superior to those of PVDF in respect to pressure sensor and shock gauge applications. Relative dielectric constant $K = \varepsilon/\varepsilon_0$ ($\varepsilon_0 = 8.85.10^{-12}$ F/m) of PVDF is 10–12 between 50 Hz and 100 kHz, and T = 25–90 °C. According to the technical data sheet, for P(VDF-TrFE) copolymer (75 %/25 % film with 50 μm (± 5 %) thickness), $K = 9.6(\pm 10$ %, 9.4(± 10 %), and 9.2(± 10 %) at 0.1, 1, and 10 kHz, and tan $\delta = 0.015$ (± 10 %), 0.016(± 10 %), and 0.032(± 10 %) at 0.1, 1, and 10 kHz.

P(VDF-TrFE) was dissolved in DMF, and the nano ring single-walled carbon nano tubes (NRCNTs) were dispersed in DMF by sonication for 1 h. The P(VDF-TrFE) solution was mixed with the NRCNT solution. The mixture solution was mechanically stirred at room temperature for 1 h and then sonicated for 1 h. The solution was poured into a 4 cm × 7 cm template for making free-standing film. The resulting composite dried in a conventional oven for overnight. For the measurement of dielectric constant of PVDF-TrFE composite with the variation of NRCNTs content, Au electrodes were deposited on the bottom and top sides by e-beam deposition as presented in Fig. 3.19.

The relative dielectric constant (K), the measure of the charge retention capacity of a medium and the dielectric loss tangent (tanδ), the angle between the capacitor's impedance vector and the negative reactive axis were measured for pristine PVDF-TrFE and those mixed with NR SWCNTs-PVP at the various frequency from 0.1 kHz to 1 MHz.

As recorded in Table 3.2, pristine P(VDF-TrFE) shows the relative dielectric constant of $K = 12.7$ at 1 kHz which is similar to that of $K = 10$–12 taken from

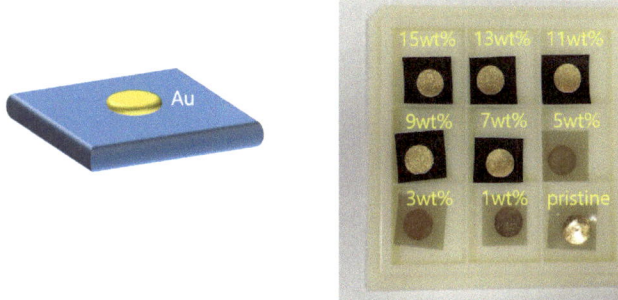

Fig. 3.19 Free standing pristine P(VDF-TrFE) and NR-SWCNTs incorporated composite films with 60–80 µm thickness

the data sheet taken from the manufacturing company. As the relative weight contents of NR SWCNTs with PVP were varied from 1 to 15 wt, K gradually increased and showed the maximum of $K = 62.9$ at 1 kHz at the relative weight of 11 % and then slightly decreased again to $K = 51.7$ at the relative weight of 15 %. K value shows huge increase amounts to 359 %. There have been several reports on the increase of P(VDF-TrFE) composite with incorporation of metal nanoparticles like Ni (Dang 2003) or Ag (Huang et al. 2009), high dielectric ceramic nano powders like $BaTiO_3$ (Channel and Jog 2008) or $Cu_3TiO_4O_{12}$(CCTO) or TiO_2 (Barber et al. 2009), hybrid oxide with metal particles like Co–ZnO (Barber et al. 2009) or Ni-CCTO (Yang et al. 2011), hybrid polymer with oxide (Fu et al. 2015) and carbon nano tube (Wang and Dang 2005). But such a high value of $K = 62.9$ in PVDF or P(VDF-TrFE) copolymer incorporated with pure organic materials, except with metal or ceramic materials, has never been reported. For dielectric materials, dielectric tangent loss (tanδ) is also a crucial factor for the application of in many electronic and electrical devices. Also it shows very low dielectric tangent loss of 0.06 at 1 kHz. After a critical relative weight amounts 13 % of incorporation, superfluous amounts of NR SWCNTs seem to deteriorate the dielectric property of P(VDF-TrFE). Such degradation is directly related to electrical breakdown. From thermogravimetric analysis (TGA), relative weight content of PVP in NR SWCNTs with PVP was measured as 23 %. Accordingly, pure amounts of NR SWCNTs showing the maximum relative dielectric constant ate equal to ca. 8.5wt%. The effective dielectric constant K of the NR-SWCNTs incorporated P(VDF-TrFE) can be calculated by percolation model or Maxwell equation or Bruggeman model based on mean field theory (Yoon et al. 2003). It needs further study (Table 3.3).

Table 3.3 Relative dielectric constant (K) and dielectric loss tangent (tanδ) with the variation of relative contents of NR SWCNTs in P(VDF-TrFE) composite

	kHz	Pristine	P(VDF-TrFE) incorporated with NR-SWCNTs-PVP							
			1 %	3 %	5 %	7 %	9 %	11 %	13 %	15 %
K	0.1	13.0	19.3	22.5	31.2	40.0	53.3	67.4	57.9	55.4
	1	12.7	18.7	21.7	29.9	38.3	50.5	62.9	54.4	51.7
	10	12.3	17.9	20.8	28.6	36.6	47.4	57.8	51.0	47.8
	50	11.6	16.9	19.6	34.4	34.4	43.7	52.5	47.3	43.8
	100	11.2	16.1	18.7	32.9	32.9	41.3	49.3	44.8	41.4
	1000	8.8	12.3	14.1	24.5	24.5	29.8	35.2	32.9	30.2
tanδ	0.1	0.01	0.03	0.04	0.18	0.20	0.24	0.18	0.47	0.36
	1	0.01	0.02	0.03	0.04	0.04	0.06	0.06	0.05	0.08
	10	0.04	0.04	0.04	0.05	0.05	0.07	0.07	0.06	0.08
	50	0.08	0.08	0.09	0.09	0.09	0.11	0.11	0.10	0.12
	100	0.10	0.11	0.12	0.12	0.12	0.14	0.14	0.13	0.15
	1000	0.19	0.21	0.22	0.24	0.24	0.25	0.24	0.24	0.26

3.8 Synthesis of Various Metal Oxide–Graphene Core–Shell Quantum Dots

Besides ZnO-nanocarbon hybrids materials, other metal oxide–graphene nanocomposites, CuO–G, SnO$_2$–G, and Mn$_2$O$_3$–G were also synthesized and analyzed by x-ray dif fraction pattern, Raman spectroscopy, and high-resolution transmission electron micro scope (HR-TEM). As shown in Fig. 3.20, all CuO–G, SnO$_2$–G, and Mn$_2$O$_3$–G quantum dots were well formed into consolidated core–shell structure and showed a quite uniform distribution of the size smaller than ~ 10 nm. As the like ZnO–G, Cu(Sn, Mn)$_2$ (CH$_3$COO)·H$_2$O) acetate hydrate were reacted with DMF and then mixed with pre-treated graphite oxide with acid.

Figure 3.21 illustrates XRD patterns of CuO–G and SnO$_2$–G hybrid quantum dots. In case of CuO–G, graphene-related XRD peaks are clearly observed at 2θ = 26° and 46° even the intensity is low. From the observation of CuO (002) and

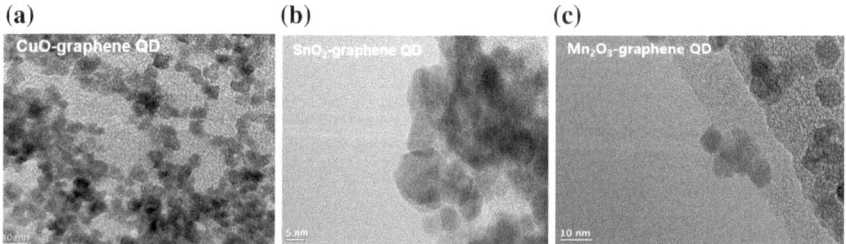

(a)   **(b)**   **(c)**

Fig. 3.20 HRTEM images of CuO–G (**a**), SnO$_2$–G (**b**), and Mn$_2$O$_3$–G quantum dots

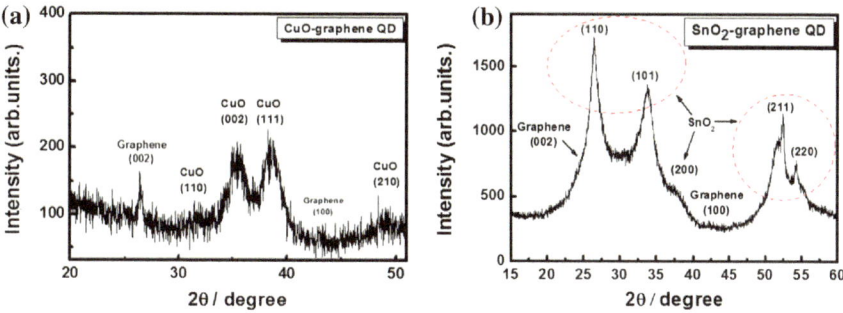

Fig. 3.21 X-ray diffraction patterns of CuO–G (**a**) and SnO$_2$–G (**b**) hybrid quantum dots

(111) peaks, it is revealed that Cu oxide crystallizes into CuO instead of Cu$_2$O. In SnO$_2$–G hybrid quantum dots, graphene peak could not be observed because the position of G(002) peak is overlapped with that of SnO$_2$(110) background.

As presented in Fig. 3.22, Raman spectroscopy in both CuO–G and SnO$_2$–G hybrid quantum dots reveals that G-peak is splitted, indicating that the attached graphene quantum dots (GQDs) are curved affected by strain. The two split are resolved into G$^-$ = 1576.6 cm^{-1} and G$^+$ = 1616.1 cm^{-1} for CuO–G and G$^-$ = 1585 cm^{-1} and G$^+$ = 1602.1 cm^{-1} (not clearly resolved) respectively. The splitting of 40 cm-1 in CuO–G is quite larger than that of 26 cm^{-1}(Son et al. 2012a, b) and thus the strain imposed on GQD in CuO–G will be larger than ε = 0.8 %.

As already introduced in Sect. 1.4, Cu$_2$O/CuO-coated CNT graphene sheet was synthesized by a polyol-mediated self-assembly method and evaluated for Li-ion battery anode (Feng et al. 2014). Hybrid CNT/SnO$_2$ has been synthesized and tested for detection of formaldehyde (Liu et al. 2008) ethanol (Chen et al. 2006), and reducing gases CO (Wu et al. 2008), NO$_2$ (Deng et al. 2005; Chrissanthopoulos et al. 2007), NH$_3$ (Hieu et al. 2008). A sol-gel coated SWCNTs with SnO$_2$ hybrid materials sensor to NO$_2$ showed considerably enhanced sensitivities compared with the pure SnO$_2$ sensor (Wei et al. 2004). The morphology and surface area of the

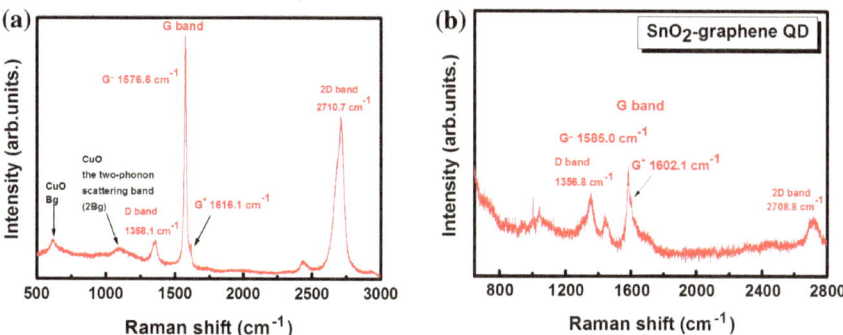

Fig. 3.22 Raman spectra for CuO–G (**a**) and SnO$_2$–G (**b**) hybrid quantum dots

hybrid sensors were not different from those of the pure SnO_2, and the observed sensitivities increased with increase of CNT loading. Therefore it can be concluded that the advanced sensing behavior originated from a common interface with CNTs. Compared to conventional SnO_2 sensors which operated at temperatures between 200 and 500 °C, the SWCNT/SnO_2 hybrid gas sensors could indeed be operated at room temperature. If the NO_2 gas molecules adsorbing on the surface of pure SnO_2, they extract electrons, leaving the oxide surface positively charged and increase the sensor resistance by the formation of depletion region. In the CNT/SnO_2 hybrid sensor, the electric properties of the oxide are strongly enhanced by the highly conducting CNTs. Consequently, the sensor resistance is dominated by the Schottky barrier at the interface between the n-type SnO_2 grains and the p-type CNTs, which causes the formation of additional depletion layers, and thereby amplifies the increase in resistance to NO_2 adsorption and enables the operation of the gas sensor at room temperature. In consequence, consolidated SnO_2 hybrid wrapped conformally with graphene will be very promising for detection of reduction gas by large enhancement of conductive graphene quantum dots.

MnO_2, one of pseudocapacitive transition-metal oxide, was used for supercapacitors, (Lang et al. 2011) because it's known for simple and scalable synthesis, high capacity for storing electrical charge, cheap cost and harmlessness to the environment. (Chang and Tsai 2003; Toupin 2004) However, the poor conductivity of MnO_2 (10^{-5} to 10^{-6} Scm^{-1}) limits the charge/discharge rate for high-power applications (Bélanger et al. 2008; Chang and Tsai 2003).

References

P.O. Anikeeva et al., Nano Lett. **7**, 2196 (2007)

P. Barber et al., Materials **2**, 1697 (2009)

A.J. Bard, L.R. Faulkner, *Electrochemical Methods: Fundamentals and Applications*, 2nd edn. (Wiley, New York, 2001)

N. Beerman et al., J. Electrochem. Soc. **147**, 2456 (2000)

D. Bélanger et al., Electrochem. Soc. Interface **17**, 49 (2008)

V. Biju et al., J. Phys. Chem. B **110**, 26068 (2006)

K.I. Bolotin, Solid State Commun. **146**, 351 (2008)

A. Cao et al., Adv. Mater. **22**, 103 (2010)

J.K. Chang, W.T. Tsai, J. Electrochem. Soc. **150**, A1333 (2003)

C.V. Channel, J.D. Jog, exPress Poly. Lett. **2**, 294 (2008)

W.D. Chemelewski et al., J. Am. Chem. Soc. **136**, 2843 (2014)

Y. Chen et al., Nanotechnology **17**, 3012 (2006)

S. Choulis et al., Adv. Funct. Mater. **16**, 1075 (2006)

A. Chrissanthopoulos et al., Thin Solid Films **515**, 8524 (2007)

O.C. Compton et al., Adv. Mater. **22**, 892 (2010)

Z.-M. Dang, Adv. Mater. **15**, 1625 (2003)

G.H. Deng et al., Carbon **43**, 1557 (2005)

F. Wang et al. 320, 206 (2008)

Y. Feng et al., AIDS Res. Hum. Retroviruses **30**, 598 (2014)

J. Fu et al., ACS Appl. Mater. Interfaces **7**, 24480 (2015)

R. Gregorio Jr., E.M. Ueno, J. Mater. Sci. **34**, 4489 (1999)
N.V. Hieu et al., Sens. Actuators **B129**, 888 (2008)
J.C. Hill, K.-S. Choi, J. Phys. Chem. C **116**, 7612 (2012)
J. Huang et al., Adv. Funct. Mater. **17**, 1966 (2007)
X. Huang et al., Appl. Phys. Lett. **95**, 242901 (2009)
P.V. Kamat, J. Am. Chem. Soc. **113**, 9705 (1991)
P.V. Kamat et al., J. Phys. Chem. **98**, 9137 (1994)
C.J. Kim et al., Chem. Mater. **17**, 3297 (2005)
H. Kim et al., Energy Environ. Sci. **8**, 247 (2015)
J.K. Kim et al., Energy Environ. Sci. **4**, 1465 (2011)
W. Kim et al., Energy Environ. Sci. **6**, 3732 (2013)
R.V.D. Krol et al., J. Mater. Chem. **18**, 2311 (2008)
A. Kudo, Y. Miseki, Chem. Soc. Rev. **38**, 253 (2009)
X. Lang et al., Nat. Nanotechnol. **6**, 232 (2011)
T.H. Lee et al., RSC Adv. **4**, 4791 (2014)
G. Li et al., Nat. Mater. **4**, 864 (2005)
B. Li et al., J. Colloid Interface Sci. **377**, 114 (2012)
D. Li et al., J. Mater. Sci. **49**, 1854 (2014a)
P. Li et al., Phys. Chem. Chem. Phys. **16**, 23792 (2014b)
J. Liu et al., J. Phys. Chem. C **112**, 6119 (2008)
L. Luan et al., Chem. Eng. J. **229**, 126 (2013)
M. Ma et al., Chem. Mater. **26**, 5592 (2014)
A. Mao et al., J. Power Sources **210**, 32 (2012)
Q. Mei et al., Chem. Commun. **46**, 7319 (2010)
B.J. Moon et al., Nano Energy **20**, 221 (2016)
A. Paracchino et al., Nat. Mater. **10**, 456 (2011)
S. Park, R.S. Ruoff, Nat. Nanotech. **7**, 217 (2009)
S.M. Park et al., Nano Lett. **9**, 72 (2009)
C.N.R. Rao, R. Voggu, Mater. Today **13**, 34 (2010)
P. Salvador, C. Gutiérrez, Surf. Sci. **124**, 38 (1983)
MdSA Sher Shah et al., Nanoscale **5**, 5093 (2013)
K. Sohn et al., Solar Energy Solar Cells **77**, 65 (2006)
D.I. Son et al., Nanotechnology **20**, 195203 (2009)
D.I. Son et al., Nat. Nanotechnol. **7**, 465 (2012a)
D.I. Son et al., Nano Res. **5**, 739 (2012b)
X. Sun, Nano Lett. **8**, 1219 (2008)
Q. Sun et al., Nat. Photonics **1**, 717 (2007)
M. Toupin, Chem. Mater. **16**, 3184 (2004)
L. Wang, Z.-M. Dang, Appl. Phys. Lett. **87**, 042903 (2005)
Y. Wang et al., Energy Environ. Sci. **4**, 2922 (2011)
B.-Y. Wei et al., Sens. Actuators **B101**, 81 (2004)
S. Woo et al., Adv. Energy Mater. **4**, 1301692 (2014)
R.-J. Wu et al., Sens. Actuators **B131**, 306 (2008)
F.-X. Xiao et al., J. Am. Chem. Soc. **136**, 1559 (2014)
H.Y. Yang et al., Org. Electron. **11**, 1313 (2010)
W. Yang et al., J. Phys. D Appl. Phys. **44**, 475305 (2011)
J. Yanh et al., Acc. Chem. Res. **46**, 1900 (2013)
D.-H. Yoon et al., Mater. Res. Bull. **38**, 765 (2003)
D.K. Zhong et al., Energy Environ. Sci. **4**, 1759 (2011)
X. Zhou et al., Appl. Surf. Sci. **258**, 6204 (2012a)
Y. Zhou et al., Science **336**, 327 (2012b)